Darwin in a New Key

DARWIN
IN A NEW KEY

Evolution and the Question of Value

William J. Meyer

CASCADE *Books* · Eugene, Oregon

DARWIN IN A NEW KEY
Evolution and the Question of Value

Copyright © 2016 William J. Meyer. All rights reserved. Except for brief quotations in critical publications or reviews, no part of this book may be reproduced in any manner without prior written permission from the publisher. Write: Permissions, Wipf and Stock Publishers, 199 W. 8th Ave., Suite 3, Eugene, OR 97401.

Cascade Books
An Imprint of Wipf and Stock Publishers
199 W. 8th Ave., Suite 3
Eugene, OR 97401
www.wipfandstock.com

ISBN 13: 978-1-4982-3119-0

Cataloging-in-Publication data:

Meyer, William J.

    Darwin in a new key : evolution and the question of value / William J. Meyer.

    xii + 124 p.; 23 cm—Includes bibliographical references and index.

    ISBN 13: 978-1-4982-3119-0

    1. Darwin, Charles, 1809–1882. 2. Darwin, Charles, 1809–1882—Criticism and interpretation. 3. Darwin, Charles, 1809–1882—Religion. 4. Evolution. 5. Teleology. I. Title.

BD581 M450 2016

Manufactured in the USA

*For Cindy and Rob
with abiding
love and gratitude*

CONTENTS

Preface | ix

1 Grandeur or Vanity? Evolution and the Question of Value | 1

2 Domestication and the Pursuit of Value:
   The Road from Domestic to Natural Selection | 16

3 Beauty and the Appreciation of Value:
   The Continuity of Life and Sexual Selection | 40

4 Human Existence: The Capacity for Understanding and
   Evaluation | 64

5 A Case for Grandeur: Value and the Evolutionary Process | 81

   Bibliography | 117

   Index | 121

PREFACE

This book stems from a long-gestating sense that our cultural conversation about the legacy of Darwin has largely overlooked an important, unresolved issue, namely, can one coherently integrate his evolutionary insights with an affirmation of the value of existence? That is, can one affirm both the gradual emergence and extinction of life and the value of life when understood in evolutionary terms? In short, is the evolutionary process ultimately one of grandeur or vanity? This is the central problem that this fresh, lean, and substantive volume seeks to address. By drawing on the philosophical perspectives of William James, Alfred North Whitehead, and others, I redirect the cultural conversation about Darwin away from the *retrospective* question of origins (how did life or species emerge?) toward the *prospective* question of value (does what evolution produces, including us, have lasting worth, or is all value ultimately lost in the deep recesses of time, and thereby evolution is just a long train to nowhere?). Based on my close reading of Darwin's autobiography and major works, I propose that the answer to this latter question is implicitly embedded in Darwin's own writings in the form of what I call a *teleology of value*, which stands in sharp contrast to a *teleology of design*. In his evolutionary outlook, Darwin rightly and explicitly rejects any form of a teleology of design (that nature was intentionally designed or prefabricated to create certain outcomes or preselected ends); but at the same time he implicitly presupposes a teleology of value (he takes for granted and indeed affirms that at least some given species, such as humans and some types of birds, pursue value in the form of beauty or other denominations of worth). By bringing to light this implicit teleology of value, my primary aim is to offer an interpretation of Darwin that enables us to integrate his view of evolution with his affirmation of the grandeur and value of nature. The question concerning the reality of God, as James perceptively recognized, is less critical for the backward-looking question of origins, which has dominated the cultural debate over Darwin, and more critical for the forward-looking question of value, in terms of making sense of the ultimate worth of all that evolution creates. Darwin, I submit, was a theist in search of a better theism and because theology had not yet caught up to him in the nineteenth century, he became increasingly agnostic and dis-integrated—caught between his mechanistic understanding of nature, on the one hand, and his affirmation of the value and beauty of the world, on the other. Whitehead's philosophy of organism, I argue, offers a way to integrate Darwin's evolutionary insights with his affirmation of the grandeur of nature. In sum, while my primary aim is interpretative, my secondary aim is constructive: specifically, to build a prima facie case for the superiority of what I call a *three-dimensional* reading of Darwin— one that recognizes and affirms the pursuit of value in nature along with survival and reproduction—over the predominant, *two-dimensional* stance of most neo-Darwinists, which reduces nature merely to a valueless machine of survival and reproduction. A three-dimensional account, instead of truncating our understanding of nature and our place within it, offers a more full-bodied perspective that enables us finally to integrate our evolutionary understanding of the world with our experience of value within it.

The title of my book is inspired by Susanne Langer's *Philosophy in a New Key*, a mid-twentieth-century work in which she too was stirred by Whitehead's philosophy to rethink some of the underlying assumptions that have shaped modern thought. In my case, in seeking to read and interpret Darwin in a new key, I have been helped by many along the way. Looking back, I appreciate the opportunities I had early on in my research to discuss some of my preliminary thoughts with engaged and attentive audiences: first, in an address to the Science and Religion Student Forum here at Maryville College in the spring of 2009 and then in an address to the Honors Society at Western Carolina University in the spring of 2010. After making significant further strides in my research, I taught a philosophy course in the spring of 2011 on Darwin and various interpretations of his work. I am grateful to my students in that course, especially to Zachary Gekas (himself now a budding philosopher), who took copious class notes and generously shared them with me and thereby helped me to capture my emerging analysis of Darwin in comparison to the predominant views of most neo-Darwinists. I began writing a first draft of this manuscript in August 2012, which I completed in December 2013. I want to thank Franklin

Preface

Gamwell, Schubert Ogden, Kevin Schilbrack, and Paul Threadgill for their close reading of and critical feedback on that complete first draft. I have spent the past few years revising the manuscript in light of their valuable feedback (and the helpful assistance of others) and in light of further reading that I have done in the field. Needless to say, whatever shortcomings remain in this work are all mine. Finally, I dedicate this work to my wife, Cindy, and to our son, Rob, both of whom continue to grace me with the gift and joy of abiding love and support.

William J. Meyer
Maryville, Tennessee
May 2015

1

GRANDEUR OR VANITY?

Evolution and the Question of Value

There is grandeur in this view of life, with its several powers, having been originally breathed into a few forms or into one; and that, whilst this planet has gone cycling on according to the fixed laws of gravity, from so simple a beginning endless forms most beautiful and most wonderful have been, and are being, evolved.

—Charles Darwin, *The Origin of Species*

Vanity of vanities, says the Teacher, vanity of vanities! All is vanity . . . I saw all the deeds that are done under the sun; and see, all is vanity and a chasing after wind.

—Ecclesiastes

One would hope that the cultural debate over Darwin is over and that his insights are fully incorporated into our twenty-first-century worldview. For more than a century and a half, that debate has focused on the question of whether Darwin's argument for natural selection accurately explains the origin of species. Creationists and intelligent-design advocates have objected in part or whole to Darwin's position because they insist it removes

God from our understanding of the world and ignores or overlooks evidence of intentional design in nature. But these objections, I judge, are misplaced. Darwin's case for natural selection is one of the most penetrating and formative contributions to the modern age. There remains, however, an important, unresolved issue that lingers below the surface, and until that issue is resolved, we will not be able fully to integrate Darwin into our twenty-first-century lives or cultural understandings. That issue pertains to the question of value. As human beings, who seek to understand and integrate all aspects of life, we cannot divorce the question of evolution from the question of worth or significance. For, as the noted biologist Jerry Coyne recognizes, evolution raises profound questions about the meaning and nature of existence.[1]

In his famous closing words to *The Origin of Species*, quoted above, Darwin intuitively senses that the descriptive question of origins begs the evaluative question of worth. Thus he concludes that there is grandeur—great value and dignity—in nature when it is understood in evolutionary terms; the evolving forms are of immense beauty, worth, and significance. But why should one conclude this? What grounds does he offer for this affirmative assessment of the evolutionary process? Like Darwin, the Teacher of Ecclesiastes takes the long view of life and offers an evaluation of it. But, unlike Darwin, the Teacher concludes that a long view, whether understood in cyclical or evolutionary terms, reveals not a world of grandeur but one of vanity and emptiness. In the long run, the Teacher might say to Darwin, all the forms that evolve under the sun become extinct, and there is nothing to which or for whom this palette of evolving and vanishing life makes any lasting or ultimate difference. Evolutionary change is like a pebble that temporarily skips along the waters of time, but in the large scheme of things, as the evolving forms become extinct, the result is just the same as if evolutionary change had never occurred at all. It all comes to naught. Hence, *evolution is just a long train to nowhere*; to think otherwise, the Teacher might say, is the height of vanity and self-deception.

One finds voices similar to the Teacher's in some contemporary discussions of Darwin and evolutionary biology. Daniel Dennett, for instance, embraces what he calls *Darwin's Dangerous Idea*, which acts as a "universal acid" that dissolves any pretense of ultimate worth or lasting value. Evolution is a mindless, aimless, and valueless process of algorithmic outcomes creating new forms that come into and eventually pass out of existence.

1. Coyne, *Why Evolution Is True*, xv, xvii.

Thus, what makes Darwin's view dangerous, according to Dennett, is to accept and fully integrate this acidic point, allowing it to seep down into and shape our existential and cultural outlook.[2] Dennett's intellectual compatriot Richard Dawkins offers a similar outlook in his book *The Selfish Gene*. According to Dawkins, all organisms, including human beings, are merely transportation vehicles and survival machines for replicating genes. But if one were to ask Dawkins what the ultimate value is of replication itself, he has no answer other than simply to reaffirm replication for its own sake, for all evaluative acts, he contends, are surreptitiously in the service of the replicating genes themselves. As he puts it, "The genes too control the behavior of their survival machines, not directly . . . but indirectly like the computer programmer."[3] Thus once again evolution is a long algorithmic or replicating train to nowhere.

In contrast to Dennett and Dawkins, who present a seemingly nihilistic outlook, Stephen Jay Gould offers a more humanistic and Kantian-like perspective, one that seeks to separate the question of value from the question of science or origins. Gould proposes what he calls the "central principle of respectful noninterference," which he labels "NOMA, or Non-Overlapping Magisteria." By proposing NOMA, Gould seeks not only to draw a distinction between the empirical questions of science and the existential, ethical, and metaphysical questions of philosophy and religion, but also, more sharply, to create a "principled and respectful separation" between them. In fact, he envisions a strict wall of separation between the discourse of science and the discourse of philosophy, value, or religion, which he refers to collectively as the "realm of human purposes, meanings, and values." So instead of viewing Darwin's idea as a universal acid that dissolves ultimate questions of value altogether, Gould seeks to separate Darwin's scientific ideas from Darwin's evaluative assessments. But this bifurcation strategy certainly does not solve the problem of fully integrating Darwin into our twenty-first-century lives or culture; in fact, it perpetuates our modern fragmented and compartmentalized outlook. For the implication of Gould's view is that, *from the perspective of science*, the evolutionary world is indeed a mindless, aimless, and valueless material process of temporarily emerging and disintegrating entities, just as Democritus suggested in the ancient world, and just as Dawkins and Dennett assert in the contemporary context. What Darwin *as scientist* offers us, Gould insists,

2. Dennett, *Darwin's Dangerous Idea*, 63.
3. Dawkins, *Selfish Gene*, 52.

is a "cold bath theory of nature," namely, the recognition that "nature is amoral," i.e., without any connection to value or worth. In contrast, what Darwin *as moralist* offers us is his personal assessment of value, specifically his claim that the evolutionary process is one of genuine grandeur and worth. But by attempting to sharply separate science from the question of value, Gould's NOMA strategy ultimately leaves both Darwin himself and our larger culture dis-integrated. Hence, to teach science is always implicitly to teach students the "cold bath" view of reality, which is that our world is without worth, value, or meaning; if one goes down the hall and teaches humanities, one can then project a positive evaluation on to this valueless nature. But the two perspectives can never be integrated into one life or outlook; a person's viewpoint is forever split into two irreconcilable perspectives: the perspective of science and the perspective of value.[4]

Over against these prevailing contemporary voices, I will argue that Darwin's view implicitly points us toward a third path that enables us to integrate his evolutionary insights with his affirmation of the grandeur of the evolutionary process. Based on my close reading of his autobiography and major works, I submit that Darwin was a theist in search of a better theism, one that could satisfactorily incorporate his evolutionary outlook. But in the nineteenth century, theology had not caught up to him, and thus he became increasingly agnostic. "In my most extreme fluctuations," he writes, "I have never been an Atheist in the sense of denying the existence of a God. I think that generally (and more and more as I grow older), but not always, that an Agnostic would be the most correct description of my state of mind."[5] Two years after Darwin published *Origin* in 1859, a fellow

---

4. Gould, *Rocks of Ages*, 5, 4, 197, 195. Gould proceeds to speak of "wisdom" (178) as the integrating third perspective (integrating science and philosophy/religion), but he never elaborates; for him, wisdom seems to be some personal, preferential, nonpublic, nonrational perspective that seeks wholeness where science and reason find none. The British philosopher Roger Scruton's argument for "cognitive dualism" appears to offer a bifurcation strategy similar to that of Gould. The world understood in scientific terms and the world understood from the perspective of human agency and interpersonal relations (the *Lebenswelt*) are, Scruton maintains, two cognitive perspectives that can never be reconciled or integrated (see Scruton, *Soul of the World*, especially chapters 2 and 3).

5. Darwin, *Autobiography*, 67. In his historical study of Darwin's life and changing view of religion, J. David Pleins notes that Darwin in his final years was quite taken with the work of William Graham. Graham described Darwin as an agnostic but "not a 'true Agnostic' inasmuch as he held 'inward conviction' about the universe—convictions the agnostic would not share in the least." Likewise, Pleins himself argues that Darwin always had a "religious side," namely, a curiosity "about religion even as he sought to develop a modern and critical take on the Bible and religion's evolution" (see Pleins, *Evolving God*,

Englishman named Alfred North Whitehead was born. Whitehead, who went on to become a noted mathematician, philosopher, and philosopher of science, sought to reconceive philosophy and religion in light of the insights of Darwin and modern science. As Whitehead remarks in *Science and the Modern World*, "It would . . . be missing the point to think that we need not trouble ourselves about the conflict between science and religion. In an intellectual age there can be no active interest which puts aside all hope of a vision of the harmony of truth." Since truth is finally one, and we live in one world, not two, Whitehead reasons, we must push toward an intellectual understanding that enables us ultimately to integrate science and the question of value. Toward this end, Whitehead argues that religion must evolve and incorporate the insights of Darwin and others. "Religion will not regain its old power," he declares, "until it can face change in the same spirit as does science. Its principles may be eternal, but the expression of those principles requires continual development."[6] Had Darwin lived another generation and read Whitehead, he would have found an evolutionary philosophy and theism that could give integrated expression to his biological and evaluative conclusions. In spite of this lack of historical overlap, I will argue that Darwin himself has the implicit makings of a more integrated view, one that can incorporate his scientific insights and evaluative outlook; indeed, he often tacitly presupposes such a view in his writings.

The key to bringing this tacit view to the surface is to offer a much-needed distinction and contrast between two very different understandings of teleology (*telos* = "aim, value, goal, purpose"), which I will call respectively a *teleology of design* and a *teleology of value*. This critical distinction, which is inspired by Whitehead, has been overlooked by the bulk of Western philosophical and scientific thinkers, including Darwin himself.[7] When modern thought has conceived of and rejected teleology, as Darwin explicitly does, it has assumed that the only possible kind of teleology is a teleology of design. By this I mean the notion that the world, with its various forms of life, is the product of some intentional and prefabricated design, either directly and immediately (creationism) or indirectly and gradually (intelligent design). A teleology of design looks *backward* to answer the

---

102–4, 109–10).

6. Whitehead, *Science and the Modern World*, 185, 189.

7. I am indebted to Franklin Gamwell for helping me to refine my distinction between a teleology of value and a teleology of design.

question concerning *how* the world or natural state of affairs *came about*. Accordingly, its usual answer points to the intentional and intelligent design of God. Such views were abundant in and prior to Darwin's day, most notably expressed in the analogical argument put forth by William Paley in 1802.[8] Paley suggested that the orderly and fine-tuned arrangements within nature resemble the order and precision of a watch, which, in turn, implies a watchmaker. Hence, if nature resembles a watch and if a watch implies the intelligent design of a watchmaker, then we can rationally infer that nature is also the result of an intelligent designer (God). It was this kind of argument that Darwin sought to discredit with his case for natural selection. But, again, precisely because the prevailing and unspoken assumption was and has been that the only possible kind of teleology is one of design, it has been widely assumed that to reject the latter is to reject all teleology as such. Hence according to Dennett, for instance, "Darwin had *reduced* [all] teleology to nonteleology."[9]

Instead of looking backward to explain how species come about, a *teleology of value* takes what is empirically given in nature, such as human beings and birds, and observes that such creatures are *forward*-looking insofar as they desire and pursue aims of value, beauty, and worth. The world we have is indeed the result of natural selection rather than prefabricated design, but the world we have is also a world in which at least some species actively pursue value in various forms, such as beauty, enjoyment, and understanding. Such activity is teleological, not in the sense of some pre-established design or preset ends, but rather in the most literal and important sense of being activity in pursuit of aims that are valued.[10] Darwin, as I will show, takes this value-pursuing activity for granted in numerous places in his writings—most notably in his discussions of domestication and of the appreciation of beauty among some species of birds, in his overall endeavor

---

8. See historical summary of Paley's work: http://www.ucmp.berkeley.edu/history/paley.html/.

9. Dennett, *Darwin's Dangerous Idea*, 65.

10. In an important essay written more than a century ago, in 1898, William James anticipated this critical distinction between retrospectively looking back to consider the question of origins and prospectively looking forward to consider the question of value or significance (James, "Philosophical Conceptions and Practical Results"). It is worth noting that the legacy of James was in some important respects an intellectual inspiration to Whitehead after Whitehead came to teach at Harvard in the 1920s. I will discuss James's essay in detail in chapter 5. Let me also note here that the contemporary philosopher John Cobb hints at, in his own way, something akin to my distinction between a teleology of value and a teleology of design (see Cobb, *Back to Darwin*, 219–20, 241).

to understand how nature works, and in his explicit evaluative assessment of the evolutionary process. Perhaps the place where he comes closest to recognizing the difference between a teleology of value and one of design is in a letter he wrote in 1860 to his friend and fellow naturalist, Harvard professor Asa Gray. In the wake of the publication of *Origin*, Gray was trying to interpret Darwin's theory of natural selection as a form of a teleology of design, which Darwin clearly rejected. In response to Gray, Darwin writes:

> I see a bird which I want for food, take my gun, and kill it, I do this *designedly*. [In contrast, a]n innocent and good man stands under a tree and is killed by a flash of lightening. Do you believe . . . that God *designedly* killed this man? Many or most persons do believe this; I can't and won't. If you believe so, do you believe that when a swallow snaps up a gnat that God designed that that particular swallow should snap up that particular gnat at that particular instant? I believe that the man and the gnat are in the same predicament. If the death of neither man nor gnat are designed, I see no good reason to believe that their *first* birth or production should be necessarily designed.[11]

At the beginning of this passage, Darwin unabashedly describes his own behavior as designedly, by which he means *teleological*; that is, he takes for granted that human beings seek and actively pursue desired aims, in this case a bird for food. As I will show in the next chapter in discussing his account of domestication, Darwin could just as easily have said here, "I see a flower which I admire and value for its beauty; I seek to grow and reproduce it purposely in order to further enjoy its loveliness." So in responding to Gray, Darwin begins by affirming that human activity is indeed teleological in the sense that it intentionally pursues aims that it values. In contrast, he proceeds to deny that the origin of the human species or any other species was the result of a teleology of design.

Thus, by recognizing that he implicitly presupposes a teleology of value while concomitantly denying a teleology of design, one can open up and offer what I call a *three-dimensional* reading of Darwin in contrast to the dominant *two-dimensional* view that defines much of contemporary neo-Darwinism. A two-dimensional reading insists that there are only two real factors at work in living organisms, namely, survival and reproduction. Any other activity, such as the appreciation of beauty, is either simply a hidden mechanism for survival or reproduction, or else is it merely epiphenomenal,

---

11. Darwin, *Autobiography*, 76–77; see also 280–81.

that is, a superfluous by-product of nature's pursuit of survival and reproductive strategies. In contrast, a three-dimensional reading of Darwin argues that in addition to survival and reproduction, at least some species actively pursue aims of value in the form of beauty or other denominations of worth. Such a view, instead of offering a truncated understanding of nature and our place within it, offers a more full-bodied account that enables us to move toward integrating our evolutionary understanding of the world with our experience and understanding of value within it. Whitehead, in his own way, nicely points toward this third-dimension by suggesting that there is a "three-fold urge: to live, to live well, [and] to live better." The "art of life," he notes, "is *first* to be alive, *secondly* to be alive in a satisfactory way, and *thirdly* to acquire an increase in satisfaction."[12] To be sure, survival and reproduction are necessary for the pursuit of value, but the desire for the increase in satisfaction is a real dimension found in nature; neither is it just a hidden mask for the other two dimensions, nor is it just epiphenomenal fluff. Furthermore, humans are most explicitly able to pursue various forms of value, but, as Darwin himself recognizes, some other creatures, such as some species of birds, also seek and enjoy experiences of value in the form of beauty. As human beings (which, if Darwin taught us anything, are part of the evolutionary process), we seek to survive and reproduce *in order to* pursue other aims, such as love, beauty, justice, enjoyment, and achievement. We survive and reproduce *for the sake of* experiencing the values of living; we do not pretend to pursue these other ends merely as hidden masks for or as empty time-fillers in between acts of survival and reproduction.

Some neo-Darwinists, such as the biologist Ernst Mayr, attempt to employ the language of teleology in a two-dimensional manner, but in doing so, they completely enervate it by not only rejecting any hint of a teleology of design but also stripping out any possible teleology of value. According to Mayr, nature is goal-directed in a mechanistic but not in a purposive way. That is to say, organisms, such as human beings, pursue ends, but not because they desire or seek their value, but rather because they are deterministically programmed toward such actions. Like Dawkins, Mayr draws on the language of computers and computer programming and characterizes all goal-directed behavior as "teleonomic," that is, as behavior programmed toward certain ends. As Mayr describes it, a "*program* might be defined as *coded or prearranged information that controls a process (or*

---

12. Whitehead, *Function of Reason*, 8.

behavior) leading it toward a given end." He proceeds to distinguish between "closed" or "genetic" programs, which are entirely encoded in the DNA, and "open" or "somatic" programs, which can learn and incorporate new information from an environment. Yet, whether the program is genetic or somatic, the important point is that there is no real desire or quest for value in this mechanistic, two-dimensional world of nature; rather, all behavior is programmed strictly toward survival and reproduction. All activity in an evolutionary world, declares Mayr, "must play a role in the survival or in the reproductive success of its bearer." In sum, "Teleonomic explanations are strictly causal and mechanistic."[13] There is no trace of value or worth here in this cold-bath view of nature.

Mayr's account illustrates the dominance of mechanistic and deterministic thinking in much of modern Western thought. As the philosopher of science Michael Ruse observes, "science is not just an exploration of the world, but an explanation through the use of metaphor. The scientific revolution of the 16th and 17th centuries was essentially a change in metaphors. One went from the Aristotelian metaphor of the world as organic to the Cartesian metaphor of the world as a machine . . . [T]oday's science," Ruse concludes, is "infused by the root metaphor of the machine."[14] Ruse's observation points to one of the great ironies of modernity, which is that modern science has tended to view nature not as a living organism but rather as a lifeless machine or as a programmed computer. (As we will see in chapter 5, Whitehead argues for a philosophy of organism over against the dominant philosophy of mechanism.) In keeping with this dominant mechanistic mind-set, Mayr's view truncates life into a complex version of Munich's famed Rathaus-Glockenspiel, the medieval clock in which wooden figures come out at preselected hours to act out life stories: they *appear* to be pursuing desired activities and values, but *in fact* they are simply preprogrammed to engage in such movement. Of course for Mayr, Dennett, Dawkins, and others, this programmed activity implies no designer but rather simply the product of mindless and aimless natural selection. But this only heightens the vanity of their conception of life and nature. What could be more empty than to be a programmed entity in a pointless evolutionary play?

There is no doubt that Darwin too was largely influenced by the dominance of mechanistic and deterministic thinking in much of modern thought. But I will argue that there is also a more humanistic and aesthetic

---

13. Mayr, *Toward a New Philosophy of Biology*, 49, 62, 54, 60 (italics original).
14. Ruse in Ansted, "Interview: Michael Ruse."

sensibility in Darwin's writings, one that implicitly affirms the genuine pursuit of value and thus the grandeur rather than the vanity of the evolutionary process. Along this line, the intellectual historian Robert Richards remarks that Darwin's "theory functioned not to suck values out of nature but to recover them" for a view of nature without a teleology of design.[15] Darwin himself, I judge, was never able to reconcile his humanistic and aesthetic inclinations (his affirmations of freedom, beauty, and value—what I call his implicit teleology of value) with his mechanistic and deterministic mind-set. Perhaps the place where he comes closest to acknowledging this unresolved tension is in another letter to Asa Gray. There, again resisting Gray's teleology of design, Darwin states, "But I know that I am in the same sort of muddle (as I have said before) as all the world seems to be in with respect to free will, yet with everything supposed to have been foreseen or preordained."[16] Whether interpreting the world naturally or theologically, Darwin did not see a way to reconcile his affirmation of freedom, on the one hand, and his assumption of determinism, on the other. Summarily put, I will argue that in order to integrate his evolutionary insights with his affirmation of the grandeur of nature, one must first allow Darwin's humanistic and aesthetic sensibilities to come to the fore and provide the lens through which one interprets his overall view of nature.

Unlike Ernst Mayr, Michael Pollan seeks to affirm the language of aim and desire in his account of nature, but, regrettably, he still does so largely through the predominant, two-dimensional perspective of neo-Darwinism. In his book *The Botany of Desire*, Pollan offers an intriguing reflection on the coevolution of plants and human desires. Often sounding like Dawkins, Pollan offers the basic thesis that the evolution of human desires, as expressed through farming, gardening, and other forms of domestication, is used surreptitiously by plants to get themselves reproduced.

---

15. Richards, *Romantic Conception of Life*, 516. My position, which suggests that one can find both mechanistic and aesthetic strands in Darwin's writings, stakes out something of a middle ground between the predominant neo-Darwinian account, which interprets Darwin as strictly mechanistic (e.g., Dennett subsumes all forms of domestic/methodical selection in Darwin under natural selection; see *Freedom Evolves*, 265), and Richards's reading of Darwin as part of German Romanticism, which appears to deny or downplay any mechanistic strands in Darwin. My contention is that both strands are found in Darwin's writings and that he himself did not reconcile the tension between them. The solution I propose, which shares some kindred spirit with Richards, is to allow Darwin's humanistic and aesthetic sensibilities to come to the fore and provide the lens through which one interprets Darwin's overall view of nature.

16. Darwin, *Autobiography*, 295.

According to this two-dimensional account, human aims, values, desires, and meanings are merely heuristic devices that plants use to disseminate their own genes. In Pollan's words,

> We automatically think of domestication, as something we do to other species, but it makes just as much sense to think of it as something certain plants and animals have done to us, a clever evolutionary strategy for advancing their own interests. The species that have spent the last ten thousand or so years figuring out how best to feed, heal, clothe, intoxicate, and otherwise delight us have made themselves some of nature's greatest success stories.[17]

We see here a truncated, two-dimensional perspective, namely, that evolution simply amounts to replication merely for replication's sake, albeit in more Dionysian fashion. Life is not really about desires, such as the desire for love, beauty, or enjoyment; or about values, such as pursuing creativity, justice, or regard for others. Rather, nature merely uses these pretenses as a subterfuge for the mindless and aimless task of replication, which fuels the long evolutionary train to nowhere.

At times, however, it should be noted that Pollan opens the door to a more three-dimensional account. First, he critiques the dominance of the mechanistic paradigm in Western thought, which has led to a global "monoculture" of "uniformity and control." Over against this, he advocates a "biological paradigm" that views nature as an organism rather than as a machine, and in making his critique, he affirms the values of diversity and complexity. Second, he recognizes that at root the issue lies in a philosophical question about our understanding of reality and our place within it. That is, he discerns that underlying our conceptions of evolutionary nature are our philosophical conceptions of the relation of the one and the many, the relation of unity and diversity. Quoting the ancient historian and philosopher Plutarch, Pollan notes that the mechanistic monoculture of our global economy "celebrates 'the One' [while] 'denying the many and abjuring multiplicity.'"[18] Third, one of his perceptive observations is to recognize the coevolutionary interrelations among species, such as among humans and plants. We indeed live in a world of interrelations, and these relations, as Darwin himself observed, have both conscious and unconscious effects. Thus, by directing the cultural conversation toward paying attention to the deep-seated philosophical assumptions in modern science, by critiquing

17. Pollan, *Botany of Desire*, xvi.
18. Ibid., 228, 224, 229.

the dominance of the machine paradigm in Western thought, by pointing toward the web of interrelations, and by highlighting the role of a teleology of desire within nature, Pollan makes some valuable contributions toward a three-dimensional understanding of evolution. Unfortunately, he himself is still largely caught in the truncated, two-dimensional web of neo-Darwinism.

So let me now begin to make a case for a three-dimensional reading of Darwin. In doing so, I will draw upon his major works (*The Origin of Species* and *The Descent of Man*), his autobiography (*The Autobiography of Charles Darwin*), and some of his other significant writings (*Variation of Animals and Plants under Domestication* and *The Expression of the Emotions in Man and Animals*). It is important to recall that Darwin himself viewed *Origin* as an "abstract," which he published once he learned that Alfred Wallace had independently come to similar conclusions.[19] In many ways, it is later works like *Descent* that contain a more complete description of Darwin's view of nature. To be sure, *Origin* is a remarkable classic, but, as an abstract focusing singularly on making a case for natural selection, it reduces or leaves out other key aspects of Darwin's account, such as his repeated discussions of beauty in *Descent*. So instead of reading his other works narrowly through the prism of *Origin*, which the predominant, two-dimensional view tends to do, I will argue that one should read *Origin* against the backdrop of his later works in order to get the full picture of Darwin's evolutionary understanding of nature. I will proceed in chapter 2 by examining his discussion of domestication in *Origin* (which is a key premise for his argument for natural selection) and by unpacking its important underlying assumptions. In chapter 3 I will turn to his many observations in *Descent* about the appreciation of beauty, especially as evidenced in some species of birds, and place his discussion of sexual selection within this context. In chapter 4 I will focus on his account of what distinguishes humans from other species, namely, our capacity for self-conscious awareness, and draw out the profound implications of this capacity. Last, in chapter 5 I will outline the philosophical underpinnings for why Darwin and the rest of us can rightly and coherently affirm the grandeur of an evolutionary view of the world. But before proceeding, let me conclude here by offering three final points: a clarification, an observation, and a response.

---

19. Darwin, *Autobiography*, 232, 224, 248; see also chaps. 11–12 on the writing and publication of *Origin*.

First, let me make fully clear the aim, scope, and audience of this work. My guiding aim is hermeneutical, that is, to offer an interpretation of Darwin that enables us to integrate his evolutionary insights with his affirmation of the grandeur of the evolutionary process. Toward this end, I propose a three-dimensional reading of Darwin, one that seeks to bring to the surface an implicit teleology of value embedded in his writings. Along the way, I will engage and contrast this three-dimensional reading with the predominant two-dimensional view of most neo-Darwinists. The two-dimensional account reads Darwin narrowly in mechanistic terms, viewing nature strictly in terms of survival and reproduction and thus categorically ruling out the possibility of any teleology of value evidenced in the world. Such an outlook, I suggest, cannot finally integrate the question of evolution with the question of value. Though I will critically engage various proponents of this two-dimensional view as I proceed and seek to build a prima facie case for the superiority of a three-dimensional reading, I do not presume here within the scope and limits of this work to offer either a full critique of a two-dimensional perspective or a full defense of my alternative proposal. A thorough critique of mechanistic thinking and a systematic articulation and defense of a teleology of value must await my future work or the work of others. In the meantime, my primary goal here is more hermeneutical than constructive. Since Darwin's influence on modern culture is both immensely broad and profoundly deep, I write this book for all persons interested in thinking seriously about whether and how one can one integrate an evolutionary understanding of the world with an affirmation of the meaning and value of existence. Thus I seek to write in a manner that is accessible to philosophers, scientists, theologians, and the wider educated public.

Second, the fact that I am a philosopher rather than a scientist has enabled me to read Darwin with fresh, open, and uninitiated eyes; it has enabled me to pay special attention to what he takes for granted and thus what he implicitly assumes and rarely if ever explicitly articulates. It is in these taken-for-granted assumptions that his underlying philosophical views are to be found. Karl Popper, in the preface to the first edition of his noted work *The Logic of Scientific Discovery*, reflects on the differences between scientists and philosophers. Scientists, he observes, are initiated into a tradition of inquiry that has "an organized structure," which is "a structure of scientific doctrines" that provides scientists with a functioning orthodoxy, i.e., with "a generally accepted" outlook on nature and how to approach problems. In contrast, "the philosopher finds himself in a different position. He does not face an organized structure, but rather something resembling a heap of ruins (though perhaps with treasure buried underneath)." As an outside philosopher rather than as an initiated scientist, I am freed up to read Darwin in a new key, in a way that unearths the "treasure buried underneath."[20] That treasure is his implicit teleology of value, which points us in the right direction for integrating his scientific insights with his evaluative assessment of the grandeur of evolution. This again is not to say that all of his assumptions fit comfortably together: his implicit teleology of value and his explicit affirmation of the grandeur of evolution sit uncomfortably with his mechanistic and deterministic understanding of nature. Nonetheless, our task is to find a way to integrate Darwin's insights with his and our evaluative assessments of the world.

Third, the contemporary philosopher Paul Sheldon Davies describes Darwin as a "shrewd rhetorician" who strategically adopts a voice of

> conversational charm that is disarming and at times alluring. On occasion Darwin raises his pitch to sing the praises of living things in Romantically charged refrains. [This] too is rhetorically effective, for Darwin's defense of the theory of evolution by natural selection, while effectively strangling to death the argument from design, is expressed in tones that sometimes verge on reverence. The news that God is dead is put in the mouth not of a madman but of a man who retains at least some sensibilities of a traditional believer...
>
> ... What I wish to suggest then is that we begin to appreciate Darwin's rhetorical insights... [By doing so,] [m]ight we succeed in convincing those with theological instincts that the right view of life is decidedly nontheological?[21]

What Davies seems to imply here is that Darwin used the language of value, beauty, and religion as part of a calculated rhetorical strategy rather than as a genuine expression of his own outlook. For instance, he implies that Darwin was not a theist in search of a better theism or even an agnostic open to religious sensibilities; rather, he was an atheist rhetorically disguised in sheep's clothing, i.e., he shrewdly adopted the soothing and soft

---

20. Popper, *Logic of Scientific Discovery*, xv. As I indicated in the preface, the title of my book is inspired by Susanne Langer's *Philosophy in a New Key*, a work in which she too is inspired by Whitehead to rethink some of the underlying assumptions that have shaped modern thought.

21. Davies, *Subjects of the World*, 3, 5.

voice of a friend of value, beauty, and religion in order to quietly announce "the news that God is dead." If this reading of Davies is correct, then it is evident that his view is sharply at odds with the one that I propose here. To briefly respond, one can undoubtedly agree with him that Darwin was a good writer who was mindful of his audience and cultural milieu. For example, in *Origin* he intentionally did not delve into the question of human origins, leaving that contentious issue for a later work (*Descent*). So it is certainly true that Darwin realized that his theory of natural selection had profound implications for religion, culture, and the place of humans in the universe. But this recognition notwithstanding, the overwhelming textual and historical evidence runs contrary to Davies's reading of Darwin as a cunning and cynical rhetorician. As I have suggested thus far and will seek to show further, Darwin took for granted the place of value in human activity, and he clearly recognized and appreciated the genuine element of beauty in nature. As his autobiography unambiguously reveals, these were not rhetorical strategies but rather genuine aspects of his outlook on the world, one that is best understood, I will argue, in three-dimensional terms. In short, as we will see in the following pages, we can take Darwin at his word: he means what he says and he says what he means.

2

DOMESTICATION AND THE PURSUIT OF VALUE

The Road from Domestic to Natural Selection

At first blush, one of the most striking aspects of *The Origin of Species* is that Darwin begins his modern masterpiece by discussing the human art and practice of domestication. In seeking to introduce and build a convincing case for the notion of evolution via natural selection, which is the guiding aim of *Origin*, Darwin takes as his starting premise the human pursuit of animal and plant variation for the sake of seeking desired human ends. Upon further inspection this focus on human activity comes as no surprise insofar as his longtime observations of domestication, his five-year journey on the *HMS Beagle*, his reading of Charles Lyell on geological change, and his reading of Thomas Malthus on economics were the key ingredients that eventually led him to the idea of natural selection. But among these, domestic selection was Darwin's methodological North Star. As he states in *Origin*, "As has always been my practice, let us seek light on this [matter] from our domestic productions."[1] In his autobiography, looking back, he further notes the importance of domestication for his research and reflections:

> After my return to England [in 1836 aboard the *Beagle*] it appeared to me that by following the example of Lyell in Geology and by collecting all facts which bore in any way on the variation of animals and plants under domestication and nature, some light

---

1. Darwin, *Origin of Species*, 155. For his discussion of the central aim of the work, see 435.

might perhaps be thrown on the whole subject. My first note-book was opened in July 1837. I worked on true Baconian principles, and without any theory collected facts on a wholesale scale, more especially with respect to domesticated productions, by printed enquiries, by conversation with skilful breeders and gardeners, and by extensive reading . . . I soon perceived that [domestic] selection was the keystone of man's success in making useful races of animals and plants. But how [this] selection could be applied to organisms living in a state of nature remained for some time a mystery to me.[2]

In *Origin* Darwin brilliantly starts from the premise of domestic selection, which he points out also inevitably has unnoticed and unintended consequences that he calls unconscious selection, and then he proceeds with key intermediate discussions of variation under nature and the Malthusian struggle for existence to introduce and lay out the case for natural selection. I will have occasion below to offer some further reflections on some of these more well-known aspects of Darwin's theory, but my principal focus will remain on the role that domestication plays in his argument and its important underlying implications.

DOMESTIC SELECTION

Let us begin by taking note of Darwin's terminology: when referring to domestication, he sometimes refers to it as "methodical selection," and at other times he refers to it as "artificial selection."[3] The former term is certainly adequate but not very precise insofar as one can employ methodical selection in a host of activities other than those devoted to domestication.[4] The latter term however is problematic, as I will seek to show below. I will contend that Darwin's understanding of domestication is best described as "domestic selection," which is a phrase that he himself regrettably appears not to use.

As I just indicated, Darwin sometimes refers to the process of domestication as "artificial selection" in order to distinguish it from natural selection. In doing so, he recognizes that the adjective *artificial* stems from

2. Darwin, *Autobiography*, 47–48.
3. For "methodical," see *Origin* 148; for "artificial," see *Origin* 153, 226.
4. For example, drafting and recruiting in athletics are certainly forms of methodical selection, but one would hope that they have not devolved to the point where we would call them forms of domestication.

the noun *artifice*, which refers to a range of human arts and skills designed to create things that are the product of purposive ends. Hence, domestication creates outcomes that are the result of human desire, aim, skill, and endeavor (i.e., the result of artifice). This is certainly the case, but, this notwithstanding, I submit that "domestic selection" is a more apt description of domestication for two reasons.

First, the term *artificial* tends today to connote that which is fake or unreal. Hence, from the vantage point of some neo-Darwinists, domestication is either simply a veiled form of natural selection, i.e., a veiled form of survival and reproductive strategies, or it is merely an epiphenomenal by-product of natural selection—a superfluous capacity generated by the real underlying processes of survival and reproduction. But if domestic selection is either merely a veiled form of natural selection or epiphenomenal, then Darwin's argument in *Origin* becomes problematic. If domestication is a veiled form of natural selection, then the argument is circular insofar as Darwin's initial premise (domestication) already presupposes his conclusion (natural selection) rather than argues for it, which in fact he does so well. If domestication is merely an epiphenomenal by-product of natural selection, then his initial premise (domestication) does not offer a convincing starting point for understanding natural selection, precisely because domestication is supposedly superfluous to nature and not a real or meaningful part of it. Put simply, why should one assume that the epiphenomenal is a clue to the natural? As far as I can see, there is no reason.

In contrast, I would venture to say that Darwin takes domestication to be a real and self-evident part of human activity and a helpful clue for understanding the processes of nature. The key point here is that Darwin takes nothing more for granted from his lived and observed experience than the fact that humans freely seek desired aims through domestication. He does not explicitly think or speak of this pursuit in terms of a teleology of value, but that in fact is what it is. As he puts it: "Man selects only for his own good," which can be defined in terms of that which not only sustains human life (survival and reproduction) but also enriches it.[5] This enrichment, this desire not only to live but also to live well and to live better, is illustrated by myriad forms of plant and animal breeding, such as flowers for beauty and dogs for companionship, as well as by countless other forms of human activity. Darwin implicitly notes this human pursuit of value on numerous occasions. Perhaps one of the more telling examples is when

---

5. Darwin, *Origin*, 132.

Domestication and the Pursuit of Value

he states that his eureka moment, in which he first clearly envisioned the theory of natural selection, came to him unexpectedly when he was pursuing the valued aim of recreational reading, which in this case was when he was reading Malthus for the sake of amusement. As Darwin writes:

> In October 1838, that is fifteen months after I had begun my systematic enquiry, I happened to read *for amusement* Malthus on *Population*, and being well prepared to appreciate the struggle for existence which everywhere goes on from long-continued observation of the habits of animals and plants, it at once struck me that under these circumstances favourable variations would tend to be preserved and unfavourable ones to be destroyed. The result of this would be the formation of new species. Here, then, I had at last got a theory by which to work.[6]

Needless to say, "systematic enquiry" is itself an aim pursued for the sake of the value of understanding, one that I will address in more detail in chapter 4. But here, Darwin describes how his pursuit of recreational reading led him to read Malthus, and, by implication, how this valued pursuit led him serendipitously to the concept of natural selection. My basic but important point is simply to note that Darwin takes for granted that he and other human beings are creatures that pursue valued aims, whether it is recreational reading, systematic enquiry, or domestic selection for the sake of breeding beautiful flowers or companionate dogs. Whatever the desired aim might be, humans pursue such aims for the sake of enjoying something that they value and not just as a veiled means for survival or reproduction. Darwin took this conclusion as an assumed and given fact.

Second, the term *artificial* tends to connote that which is outside of or contrary to nature, but domestic selection is neither of these precisely because *the capacity* to pursue desired aims is, by Darwin's own account, actualized through the evolutionary process. Human activity does not float above or stand outside of nature; it is itself part of nature. Hence, what humans do is not artificial but rather an expression of a natural capacity to pursue desired aims and valued ends, such as the value of novelty. As Darwin puts it, "it is in human nature to value any novelty, however slight."[7] Furthermore, as we will see in the next chapter, this capacity is also exemplified in some other species, albeit to a lesser degree. What is required, therefore, is a more comprehensive understanding of nature that can make

---

6. Darwin, *Autobiography*, 48 (italics added).
7. Darwin, *Origin*, 97.

full sense of domestication rather than treating it as something artificial or explaining it away. Because Darwin lacked a clear understanding of the difference between a teleology of design and a teleology of value, he failed to provide this more comprehensive understanding; yet as I will show, his affirmation of domestication presupposes it. So, in sum, the phrase "domestic selection" accurately describes the selective activity of domestication without the misguided connotations that "artificial selection" entails. Hence, I will use the former instead of the latter.

With these remarks in place, let us now delve into Darwin's discussion of domestic selection in order to see how he implicitly takes for granted a teleology of value. As stated above, he seeks to build his case for natural selection by first carefully analyzing the practice of domestication. "In man's methodical selection," he observes, "a breeder selects for some definite object." That is, because humans value some things more than others, because they make evaluative judgments of greater and lesser worth, they act in accord with and in pursuit of these desired aims. In the case of domestication, this means that because they value some specific trait, they intentionally seek to breed or develop that characteristic. As Darwin remarks: "One of the most remarkable features in our domesticated races [of plants and animals] is that we see in them adaptation, not indeed to the animal's or plant's own good, but [rather] to man's use or fancy." Domestic selection is an art for the sake of pursuing valued human ends. But what makes this art particularly powerful, adds Darwin, is its accumulative effects. "The key is man's power of accumulative selection: nature gives successive variations; man adds them up in certain [desired and intended] directions useful to him." Likewise, Darwin states: "Man can and does select the variations given to him by nature, and thus accumulate them in any desired manner. He thus adapts animals and plants for his own benefit or pleasure."[8] Darwin makes this same point in his later work devoted to domestication. There, he writes: "Man can select, preserve, and accumulate the variations given to him by the hand of nature almost in any way which he chooses... Man may select and preserve each successive variation, with the distinct intention of improving and altering a breed, in accordance with a preconceived idea [and aim]."[9] Back in *Origin*, Darwin further illustrates this power of desired and accumulative selection in the case of flowers: "We see an astonishing improvement in many florists' flowers, when the flowers of the present day

---

8. Ibid., 148, 89, 90, 441.
9. Darwin, *Variation of Animals and Plants under Domestication*, vol. 1, 3 [3/4].

are compared with drawings made only twenty or thirty years ago." Such intentional pursuit of accumulative ends, Darwin observes, leads to "the increased size and beauty" of whichever flowers one values.[10]

If these examples of domestication are not illustrative of and evidence for a teleology of value, then I do not know what would be. Darwin rightly took it for granted that humans are creatures that seek aims that they value, such as the beauty of a red rose, and that they use methodical means to achieve their desired ends. To be sure, domestication is often used as a means of feeding and clothing human beings for the sake of sustenance and survival. But there is no doubt in Darwin's mind, nor should there be in ours, that a great deal of domestic selection is pursued for the sake of aims that are desired and valued for their own sake, such as beauty, sport, or companionship. Humans are valuing creatures by nature; beyond the essentials of survival and reproduction, they seek aims that they desire in order to add value and enrichment to their individual and collective lives. Again, as Darwin himself states, "It is in human nature to value any novelty, however slight."[11] Domestic selection is just one creative way that humans pursue valued ends, such as novelty and beauty.

UNCONSCIOUS SELECTION AND THE TRANSITION TO NATURAL SELECTION

Having established domestic selection as a powerful and widely evident first premise, Darwin proceeds to note how this intentional and systematic pursuit has unwitting and unintentional consequences, which he calls "unconscious selection." That is, by intentionally and methodically seeking to reproduce that which is valued, breeders always choose those individual plants or animals that have the desired characteristics. But by so doing, they at the same time do not choose to reproduce those that lack that desired characteristic. Hence, without intending or being aware of it, over time they gradually begin to change the breed itself. This is what Darwin means by "unconscious selection," which he sees as a common if not inevitable by-product of domestic selection. He writes:

> But, for our purpose, a kind of Selection, which may be called Unconscious, and which results from every one trying to possess and

---

10. Darwin, *Origin*, 91, 95.
11. Ibid., 97.

> breed from the best individual animals, is more important. Thus, a man who intends keeping pointers naturally tries to get as good dogs as he can, and afterward breeds from his own best dogs, but he has no wish or expectation of permanently altering the breed. Nevertheless I cannot doubt that this process, continued during centuries, would improve and modify any breed . . . Slow and insensible changes of this kind could never be recognized unless actual measurements or careful drawings of the breeds in question had been made long ago, which might serve for comparison.[12]

As the desired characteristics are repeatedly selected and reproduced in an ongoing way, and the undesired characteristics are continuously left behind, the strong accumulative effects change the nature of the breed as a whole: "As soon as the points of value of the new [desired] sub-breed are once fully acknowledged, the principle, as I have called it, of unconscious selection will always tend . . . slowly to add to the characteristic features of the breed, whatever they may be."[13] So characteristics that were originally just a desired variation or subset of the breed now come to be common to the breed as a whole through a gradual unconscious process of change. "It can, [thus], be clearly shown that man, without any intention or thought of improving the breed, by preserving in each successive generation the individuals which he prized most, and by destroying [or neglecting] the worthless individuals, slowly, though surely, induces great changes [in the breed itself]."[14]

The genius of this second premise is that Darwin has now demonstrated that significant accumulative changes can occur in nature without aim or intention. He is thus slowly but surely moving the reader into position to recognize how such accumulative changes indeed occur on a grand scale over long periods of time via natural selection. But in moving toward this central claim, one should never lose sight of the fact that Darwin himself has already clearly affirmed the reality of domestic selection, with its intentional pursuit of value. Unlike Darwin, too many contemporary readers lose sight of this key starting point in their hurried race to natural selection.

Before reaching his famous conclusion, Darwin recognizes that there are two other essential premises, namely, variation within nature and the struggle for existence. He takes up the first of these in chapter 2 of *Origin*.

---

12. Ibid., 93.
13. Ibid., 97–98.
14. Darwin, *Variation of Animals and Plants*, vol. 1, 3 [3/4].

Though he did not have the benefit of knowing the later impact of Mendelian genetics, and thus he was not in a position to accurately explain inheritability or how nature generates variety, Darwin, like other naturalists, certainly recognized the countless variety that nature produces within any given type of plant or animal. It is this natural variety, he observes, that provides domestic selection with a palette of possibilities from which to choose, and which to methodically develop. Darwin spends much of chapter 2 outlining his antiessentialist and nominalist view of the concept of species. That is, Darwin does not draw a sharp line between *species* and *varieties*. In his words, "I look at the term species," he says, "as one arbitrarily given for the sake of convenience to a set of individuals closely resembling each other, and that it does not essentially differ from the term variety, which is given to less distinct and more fluctuating forms."[15] Countering the legacy of the teleology of design and creationist perspectives, which viewed different kinds of plants and animals as distinct, fixed, and essential types (i.e., as fundamentally different species), Darwin instead argues that "species" are merely labels that we use for the sake of convenience to describe sizeable aggregates of individuals that show a close resemblance to each other. So the difference between "species" and mere "varieties" is a matter of degree not kind. To make this determination, Darwin appeals to the informed judgment of experienced and competent judges: "Hence, in determining whether a form should be ranked as a species or a variety, the opinion of naturalists having sound judgement and wide experience seems the only guide to follow. We must, however, in many cases, decide by a majority of naturalists, for few well-marked and well-known varieties can be named which have not been ranked as species by at least some competent judges."[16]

As noted above, the final catalyst propelling Darwin to the concept of natural selection was his reading of Malthus's economic treatise, which sets the conceptual framework for chapter 3 of *Origin*. According to Malthus, human population tends to grow geometrically while food production grows merely arithmetically; this critical difference leads to more people than food and thus, over time, leads to human strife and starvation. Darwin takes this principle and applies it to all forms of life within nature. In his words:

15. Darwin, *Origin*, 108.
16. Ibid., 104.

> *A struggle for existence inevitably follows from the high rate at which all organic beings tend to increase.* Every being, which during its natural lifetime produces several eggs or seeds, must suffer destruction during some period of its life, and during some season or occasional year, otherwise, on the principle of geometrical increase, its numbers would quickly become so inordinately great that no country could support the product. Hence, as more individuals are produced than can possibly survive, there must in every case be a struggle for existence, either one individual with another of the same species, or with the individuals of distinct species, or with the physical conditions of life. It is the doctrine of Malthus applied with manifold force to the whole animal and vegetable kingdoms; *for in this case there can be no artificial increase of food, and no prudential restraint from marriage.* Although some species may be now increasing, more or less rapidly, in numbers, all cannot do so, for the world would not hold them.
>
> *There is no exception to the rule that every organic being naturally increases at so high a rate*, that if not destroyed, the earth would soon be covered by the progeny of a single pair. Even slow-breeding man has doubled in twenty-five years, and at this rate, in a few thousand years, there would literally not be standing room for his progeny.[17]

What this Malthusian principle does for Darwin's argument is that it posits fundamental conditions of pressure within nature, combined with nature's own generation of variety and fecundity, to create conditions of struggle leading inevitably to winners (those that survive and reproduce under these conditions of pressure) and losers (those that perish or gradually diminish). Having already demonstrated the effects of unconscious selection and the variations produced in nature, and having here demonstrated that the conditions of struggle lead to inevitable winners and losers, Darwin understands the concept of natural selection as merely the last logical step in the argument to articulate the inevitable implications of this natural winnowing process. But before getting to that, let me briefly reflect on Darwin's use of Malthus.

It has always been somewhat unclear to me whether Darwin thinks human population growth always proceeds at a geometrically high rate or that it does so only sometimes. In the extended passage above, I have used italics to highlight some key lines. In the final paragraph, he states: "There is no exception to the rule that every organic being naturally increases at so

---

17. Ibid., 116–17 (italics added).

high a rate." This certainly seems to suggest that humans, like every other organic being, always increase at such a pace. He follows this up by declaring that "even slow-breeding man has doubled in twenty-five years, and at this rate, in a few thousand years, there would literally not be standing room for his progeny." This illustration seems to reinforce the notion that human reproduction always occurs geometrically. In contrast to these seemingly unequivocal statements, the initial line of the passage appears to open a window in a different direction: "A struggle for existence inevitably follows," he says, "from the high rate at which all organic beings *tend* to increase." Does Darwin mean here that all species *tend* toward geometric growth but that environmental factors, such as drought, can impede this tendency and, moreover, in some species, such as humans, this tendency need not always be pursued? That is, in the human species, which Darwin elsewhere acknowledges has a higher degree of self-conscious awareness (a point I will develop in chapter 4), this capacity enables humans to pursue other aims and values and thus enables them to limit their reproductive tendencies in the pursuit of these other ends.[18] Notice, for instance, in the middle of the passage where he precludes from his application of the Malthusian principle to the plant and animal kingdoms any possibility of an "artificial increase of food, and no prudential restraint from marriage." I take this to imply that in the case of human beings, unlike plants and other animals, there can be a lessening of population pressure because of the art and skill of domestic selection and agriculture (artifice), and because of the capacity to willfully limit reproduction. Indeed, this appears what Darwin does in fact mean and certainly ought to mean given his prior discussion of domestic selection. If contemporary China's controversial "one-child policy" illustrates anything, it is that humans can indeed intentionally seek to control and limit their reproduction. Furthermore, this human capacity to limit reproduction is indicative not only of virtuous or coercive prudence—of being able to limit population *for the sake* of future survivability—but indicative also of the fact that humans seek to pursue other values besides survival and reproduction and thus choose fewer offspring *for the sake* of pursuing these other goods. The relatively low birthrates in contemporary North America and northern Europe, for instance, are indicative of human beings choosing to have smaller families or no children at all for the sake of seeking other valued ends, such as education, professional fulfillment,

---

18. Darwin, *Descent of Man*, 105.

freedom, recreation, travel, and so forth.[19] Thus, it is evident that human reproduction need not be under the shadow of population pressure.

NATURAL SELECTION

It is precisely this population pressure, however, that is a necessary condition for natural selection. In chapter 4 of *Origin*, Darwin makes a powerful and rationally convincing case that natural selection is the inevitable outcome of variation in nature *plus* population pressure *leading to* a struggle for existence, which in turn *inevitably leads* over long periods of time to natural selection. He states:

> Can it, then, be thought improbable, seeing that variations useful to man have undoubtedly occurred [through domestic selection], that other variations useful in some way to each being in the great and complex battle of life, should sometimes occur in the course of thousands of generations? If such do occur, can we doubt (remembering many more individuals are born than can possibly survive) that individuals having any advantage, however slight, over others, would have the best chance of surviving and of procreating their kind? On the other hand, we may feel sure that any variation in the least degree injurious [or disadvantageous] would be rigidly destroyed. This preservation of favourable variations and the rejection of injurious variations, I call Natural Selection.[20]

Darwin's insight is to recognize the rational implication of this almost equation-like relation: natural variation + reproductive pressure → struggle for existence → natural selection. Indeed, Darwin might just as well have called it the principle of *natural implication*, indicating the inevitable winnowing implications of certain factors and conditions in relation to one another. Along these lines, the philosopher Christian Illies describes natural selection as "Darwin's a priori insight." Darwin's "ingenious and enormously productive insight," observes Illies,

> is centrally one argument in four steps. Roughly we can state them as follows: (1) Animals or plants produce more offspring than can survive, since the resources they need are limited. The consequence is the struggle for existence. (2) The members of a species always differ in their features to a certain degree; some are better

---

19. See Taha, "Opting Out of Parenthood."
20. Darwin, *Origin*, 130–31.

adapted to their environment than others. These will have advantages and thus be positively selected in the struggle for existence. They will gradually replace less adapted ones. (3) Over time the selection of minor variations will accumulate and a new species will develop. (4) If the selection happens in different ways, because the species inhabits an area with a divergent environment, then a parental species can branch into different new ones.[21]

In a later work, Darwin himself acknowledged that "the term 'Natural Selection' is in some respects a bad one, as it seems to imply conscious choice," which is clearly not what he means.[22] Rather, his perspicacity is to discern how a winnowing process inevitably occurs in nature purely by rational implication when all the right factors are in place. There is no teleology of design here precisely because the winnowing process is not an active pursuit of anything let alone any aim. It is simply an inevitable by-product or implication of what happens when certain factors overlap. This winnowing process is like sand falling through an hourglass: no one today would say that the sand seeks to fill the bottom; rather, it is simply the inevitable implication of sand falling through a narrow opening. Likewise, natural selection is not the pursuit of any aim, not even seeking the most fit; rather, those organisms most adapted to a given environment are those, by implication, that have the best chance over time of reproducing and continuing on in that environment. But, of course, it was from the North Star of domestic selection that Darwin took his methodological bearings and from which he gained a vantage point to recognize this winnowing process at work. Hence, he defends his use of the phrase *natural selection*: "The term is so far a good one as it brings into connection the production of domestic races by man's power of selection, and the natural preservation of varieties and species in a state of nature."[23] The key connection, I judge, is found in the power of accumulative effects over time: in the case of domestic selection, Darwin notices how a desired quality is intentionally enhanced over time, and how this pursuit of a valued aim also has unintentional accumulative consequences on the breed as a whole, i.e., unconscious selection; in the case of natural selection, he infers how, given natural variation and a struggle for existence, enormous accumulative effects occur inevitably by implication over vast periods of time. Darwin

---

21. Illies, "Darwin's A Priori Insight," 63.
22. Darwin, *Variation of Animals and Plants*, vol. 1, 5 [5/7].
23. Ibid.

summarizes these interrelated elements and their inevitable implications by clearly noting that he would never have understood how modification is effected in nature "had I not studied domestic productions, and thus acquired a just idea of the power of selection. As soon as I had fully realized this idea, I saw, on reading Malthus on population, that Natural Selection was *the inevitable result* of the rapid increase of all organic beings; for I was prepared to appreciate the Struggle for Existence by having long studied the habits of animals."[24]

REFLECTIONS ON DOMESTIC AND NATURAL SELECTION

With this outline of Darwin's discursive path from domestic to natural selection now in place, let me conclude by offering a set of three detailed reflections on these important matters.

First Reflection

It is evident that one certainly need not affirm or presuppose a teleology of design in order to recognize and affirm a teleology of value, as Darwin's own essential use and discussion of domestication clearly illustrates. Darwin's American friend Asa Gray tried to interpret Darwin's theory as an incremental and veiled form of design, but Darwin rightly and unequivocally rejected all such interpretations, even at the same time as his theistic or at least agnostic inclinations were not hostile to such perspectives. He notes: "However much we may wish it, we can hardly follow Professor Asa Gray in his belief 'that [natural] variation has been led along certain beneficial lines,' like a stream 'along definite and useful lines of irrigation.' If we assume that each particular variation was from the beginning of all time preordained," then the clear evidence of all the variations that are not beneficial, plus the evidence for geometric reproduction, which leads to a struggle for existence and, in turn, inevitably leads to natural selection, "must [all] appear to us [as merely] superfluous laws of nature. On the other hand, an omnipotent and omniscient Creator ordains everything and foresees everything. Thus we are brought face to face with a difficulty as insoluble as is that of free will and predestination."[25] One sees here that

---

24. Ibid., 8–9 [10/11] (italics added).
25. Darwin, *Variation of Animals and Plants*, vol. 2, 371–72 [428].

Darwin's thinking had clearly outpaced his theological inheritance. He is in need of a better theism, one that critically revises or abandons traditional notions of omnipotence, omniscience, and divine determinism, in order to make full sense of his joint affirmations of domestic and natural selection. On one side, he clearly affirms that human beings freely pursue valued aims and goods through the methodical means of domestic selection. As he acknowledges, he would never have recognized the accumulative power of selection in nature without his careful study of domestication. On the other side, he concludes that the origin of any species, including humans, came about through a process of natural implication, not through a prefabricated design. So he needs a more comprehensive perspective that coherently affirms *that* at least some evolutionary creatures explicitly pursue value, meaning, and worth through their choices and methodical selections while, at the same time, affirming that *how* such creatures evolved was not itself a product of intentional or preordained design. In short, Darwin's affirmation of domestic *and* natural selection calls for a teleology of value without subscribing to a teleology of design.

Second Reflection

Someone like Ernst Mayr might try to counter my notion of a teleology of value by suggesting that domestication is not really the free pursuit of desired ends but rather is a determined and preprogrammed response in the human organism; it is not a goal-directed teleology of value in which humans seek aims that they desire but rather a "teleonomic form [of activity] . . . governed by a program [either directly or indirectly]." As we saw in chapter 1, Mayr seeks to maintain the language of goal-directed behavior in nature by transforming it into a preprogrammed response. "A teleonomic process or behavior is one which owes its goal-directedness to the operation of a program . . . [A] 'program' exists which is causally responsible for the teleonomic nature of a goal-directed process."[26] It is true that at one point Mayr appears to equivocate as to whether his deterministic account applies to human activity. He declares: "Intentional, purposeful human behavior, is almost by definition, teleological. Yet I shall exclude it from further discussion because use of the words *intentional* or *consciously premeditated* . . . runs the risk of getting us involved in complex controversies over psychological theory, even though much of human behavior does not differ in kind

---

26. Mayr, *Toward a New Philosophy of Biology*, 55, 45.

from animal behavior."[27] If Mayr were to exclude human behavior from his overall account, he would imply some form of dualism, which certainly runs counter to Darwin's whole point about the inclusion of humans within our understanding of nature. Thus, it is the last part of this quoted passage that is most telling where Mayr insists that most "human behavior does not differ in kind from animal behavior." In short, Mayr appears to stick to his central claim, which is that what appears to be purposive, value-pursuing, future-directed behavior is really in fact strictly causal and mechanistic activity wholly determined by the past. So human activity, which *appears* to be the free pursuit of desired ends, such as illustrated by domestic selection, is *really* in fact the playing out of a preprogrammed set of movements. Instead of a teleology of value striving toward the realization of some desired future state of affairs, what is really going on is a deterministic unfolding or push from the past by some programmed activity that has no genuine relation to any pursuit of desire or value.

Let me offer multiple lines of response to this view. To begin with, if Mayr or others want to hitch Darwin's evolutionary wagon to a deterministic metaphysic, then we can have that philosophical debate—and it is a *philosophical* not a biological debate. Darwin himself gives no hint or indication that he believes domestic selection is a programmed activity rather than what it clearly appears to be, namely, the free pursuit of desired aims through systematic means. Furthermore, pragmatically speaking, I have yet to meet any determinist who actually lives in accordance with his or her own stated conviction. As Charles Peirce and the other pragmatists note, the meaning of a proposition is rooted in its present and future application "to human conduct," and "different beliefs are distinguished by the different modes of action to which they give rise."[28] And, yet, neither Mayr nor anyone else appears to live his or her life as if they believe that what they do is in fact a programmed response rather than a free pursuit of valued aims; put simply, their stated belief in determinism does not appear to give rise to a different mode of action within their own lives. Let me offer one detailed example.

In his book *Subjects of the World: Darwin's Rhetoric and the Study of Agency in Nature*, which I briefly addressed at the end of chapter 1, Paul Sheldon Davies goes to great lengths to undercut our perceived notions of human freedom, responsibility, teleology, and subjectivity, which are due in

---

27. Ibid., 41.
28. Peirce, *Selected Writings*, 194, 121.

part, he contends, to psychologically generated and satisfying illusions and in part to the religious and theological baggage of our cultural inheritance. Davies seeks to overturn not only some versions of human agency, such as those that affirm an unchanging soul, but any and all notions that affirm our first-person, lived sense of pursuing aims and values. "We should expect that what feels or appears important to us," remarks Davies, "is often not important... [M]any of our most important concepts—those concerning human agency, in particular—do not conform to the way the world in fact is." Davies argues that the results of recent neuroscience strongly suggest that our sense of self is mostly if not totally illusory. Our human actions are largely if not wholly determined by natural bodily mechanisms and psychological interactive subsystems far below the radar of conscious awareness; we are the products of natural selection, and our actions continually bear that out in all things, even if we do not consciously recognize it or believe it. Thus, what we consciously think we are doing, such as pursuing values of love, beauty, knowledge, and justice, is all just part of the lie that we tell ourselves. To borrow language from aviation, we think we are flying our lives manually toward desired ends, but, in fact, our lives are mainly if not wholly flown on autopilot without us even knowing it. The only true way, therefore, to know human beings, contends Davies, is from the third-person perspective of science, which reveals these unconscious, deterministic mechanisms at work, not from the first-person perspective of lived experience. In essence, what science reveals is that the whole is less than the parts insofar as it is really the parts that are driving the whole, not the whole integrating and directing the parts. Hence, according to Davies, we are "Subjects of the World" in a twofold sense: first, we are deterministically subject *to* nature (we are subjects of nature, not freely acting citizens of the world); and, second, we are subjects *of* scientific research and inquiry, since we are only truly understood from the third-person perspective of science. What we are definitely *not* are agents of freedom and subjectivity, consciously pursuing aims and values in the world; this is the predominant misconception that science can help us to eliminate, if only we have the cultural courage and persistence to allow it. "Whether or not we care to face it," Davies declares, "we are subjects of the natural world... That we can know what we are like prior to actual [scientific] inquiry is a particularly pernicious lie that we, thanks to the very structure of our psychology, are geared to tell ourselves. It is also a lie we are geared to believe."[29]

29. Davies, *Subjects of the World*, 33, 34. For a good overview of his position, see

Having carefully read Davies's extended argument over 226 pages, what I found most striking was the picture of him lovingly embracing his young daughter on the inside fold of the dust jacket at the back of the book. After having systematically sought to undercut the basis of our lived experience of agency and value, there is Davies willingly, genuinely, and smilingly holding his daughter in a loving embrace. According to the implications of his stated thesis, his actions are not really about love and the affirmation of worth (that is the lie we tell ourselves) but rather are the product of deterministic mechanisms below the surface. And yet, my pragmatic point is that Davies does not really believe his own thesis when it comes to his own life, relationships, and actions; in short, his photo tells a quite different story than his text does, namely, that he lives and loves as an acting agent pursuing value in the world, and not as one who genuinely believes he is determined by or merely subject to the world. Moreover, at the front of the book, he dedicates the work to this same daughter with the words: *"For Cassandra Cailin my Daughter Subject of the World more than imagination can Embrace or Bear to Behold."*[30] Surely these words are meant to convey his love and affirmation of her, that there is more to her than science can ever behold, and that she herself is neither merely a subject of the world in a deterministic sense nor a subject of scientific observation, but rather that she is in fact a valued and loved individual, an agent, a young subjectivity beginning to live and direct a life worthy of respect, dignity, and awe. Indeed, on the book's very last page of text, in a separate acknowledgement section, Davies admits this split between his stated theory and his lived experience. He suggests that in our personal relations, "a feeble imagination," by which he appears to mean an *intellectual blinder*, is needed in order to help "us avoid various sorts of trouble. Most every father and mother knows," he reflects,

> that becoming a parent creates vulnerabilities that can crush the imagination. Certain thoughts become unthinkable, or at least unbearable. And in my own case, being the father to such startling intelligence and beauty, and to a convulsive excitement for the world that her growing body just barely contains, is, at times,

---

19–34 and 224–26.

30. Ibid., dedication page. Though my wording is exact, for the sake of space and style I do not break up the dedication into separate lines and do not use block capital letters, which is how it appears in the book; instead, I use italics to convey the personal quality of the dedication.

> simply too much. An infirm imagination, thank goodness, protects me from what I know.
>
> Love sometimes enfeebles the imagination and pushes knowledge away; love and knowledge do not always get along. We are emotional and cognitive aggregates cast from a merciless history that cares nothing for integration. Our constitutional conflicts even include the naïve conviction that we are free of such conflicts. That is the kind of agent we are. The only consolation—if that is what this is—is that at certain moments, if we are lucky, we get to know and understand a few truths about the world, and then, at other moments, if we are lucky, our imaginations [i.e., knowledge] go to sleep and we, as fathers and mothers, get lost in the convulsing intelligence and beauty of new and ascending life.[31]

In this revealing passage, Davies, like Dennett, takes the legacy of Darwin and contemporary neuroscience to be a universal acid that dissolves any notion of ultimate worth or integration and any deep bond between value (love) and knowledge. On the one hand, he seems to want to suggest that he does believe his own theory in his personal life, but, on the other hand, he indicates that he has to get intellectually drunk, so to speak, in order to forget it for a time in order to lose himself in the love of his daughter and others. Since we are mere bundles or "aggregates" of sundry thoughts and feelings in a valueless universe, he suggests, the best we can hope for is to learn a few truths about ourselves and the world, and then, "if we are lucky," to be able to allow our knowledge to "go to sleep" for a period of time in order to allow us "to get lost in the . . . beauty of new and ascending life." Affectively, Davies believes in the reality of love and the value of beauty, but cognitively he does not; thus, he acknowledges that he needs to set aside his theory in order to live out the most important parts of his personal life. Hence, I reiterate my pragmatic point, which is that neither Davies nor Mayr nor anyone else lives their life in accordance with a belief in determinism.

On another but related matter, it should be noted how Davies places such total and uncritical trust in scientific observation but, at the same time, totally distrusts any first-person lived experience, including by implication the pursuit of value in the form of scientific understanding itself. Pursuing science, as Davies wholeheartedly recommends, is itself the aim, goal, and telos of some individuals living and acting in the world. It is *agents* who observe, measure, interpret, and assess the data of their experience; data must

31. Ibid., 241.

always be interpreted by some agent as part of that agent's lived experience: the data itself may come from micro- or macroscopic technology, but the act of perceiving, interpreting, and assessing the data is inescapably part of the individual's first-person lived experience. It is *I* who see, interpret, and assess the brain-imaging scan; the sophisticated devices and equipment are tools that we have developed *for the aim or goal* of pursuing scientific understanding. Davies, like Whitehead (as we will see in chapter 5), is indeed right to insist that the act of seeing and consciously reflecting is a process that involves the influence and function of various bodily parts and neurological elements, but the culmination of that process is an individual or agent, a whole that lives and acts in the world as an integrated self who pursues aims and values, such as scientific understanding.[32] Thus, if we do not trust our own everyday experience of being individuals freely pursuing desired aims and values, such as pursuing scientific inquiry, and if we do not trust our observations of others pursuing aims in their own lives, then why should Davies, Darwin, or we ever trust our empirical observations of humans or the rest of nature? In sum, if the evidence from our lived experience and from our first-person empirical observations cannot be largely trusted, then scientific inquiry is dead.

My final line of response to Mayr is to raise the issue of the genetic fallacy. Just because the human species evolved through the winnowing process of natural selection does not mean that *all* human behavior is merely behavior about survival or reproduction. Potentiality that was inchoate, latent, or switched off at earlier phases of evolutionary development can become manifest at later stages. Hence, the question of origins, the question of *how* something came to be, does not answer the question of *what* something in fact is: to think otherwise is to commit the genetic fallacy. Just because we all originated as infants does not mean that all that we do as adults is really just infantile behavior in disguise, Freud notwithstanding. Likewise, just because the capacity to pursue value in a conscious and intentional manner came to fruition via natural selection does not mean that such pursuit of value is simply programmed survival or reproductive strategies in disguise.

---

32. See Whitehead, *Adventures of Ideas*, 46–48, for an illuminating discussion of human activity that touches upon similarities and differences between Whitehead's view and that of a thinker like Davies. Also, Roger Scruton in his book *The Soul of the World* seems to make two points similar to those I am making here: namely, that it is an agent or *I* who interprets the data (65–66) and that "consciousness is a property of... the whole person" (62).

Domestication and the Pursuit of Value

Third Reflection

My final reflection focuses on the contemporary debate among neo-Darwinists concerning whether the actual outcome of evolution via natural selection was the result of massive contingencies, and thus would not have duplicated itself if the process played out again; or whether evolution involves the playing out of certain inevitable options that would always lead to similar outcomes. Put simply, is it likely that humans or any other particular species would evolve again if the evolutionary process were carried out a second or third time? To be clear, it appears that both sides agree with Darwin that natural selection *itself* is an inevitable process given the conditions of variation, overpopulation, and a struggle for existence. Thus the debate is certainly not over whether natural selection is at work. Rather, the debate is over the ingredients and results of the process. To use a rough analogy from the kitchen, both sides agree that a blender will always blend (natural selection), but they disagree over whether the same ingredients would go into the blender on repeated occasions and, thus, whether the outcomes would be similar. On one side stands Stephen Jay Gould, who argues for massive contingency and nonrepeatability. That is, he insists on the "theme of radical contingency and [the] improbability for particulars, whatever the predictability of general patterns (with humans clearly defined as an improbable particular)." He adds, "radical contingency is a fractal principle, prevailing at all scales with great force. At any of a hundred thousand steps in the particular sequence that actually led to modern humans, a tiny and perfectly plausible variation would have produced a different outcome, making history cascade down another pathway that could never have led to *Homo sapiens*, or to any self-conscious creature." Hence, we humans "are glorious accidents of an unpredictable process with no drive to complexity."[33] For Gould, humans are a statistical outlier in an evolutionary world that has no teleology of design and no urge toward complexity or creativity.

In contrast, others, such as Daniel Dennett and Simon Conway Morris, though coming from very different metaphysical outlooks, argue that evolution via natural selection would lead on multiple occasions to similar outcomes because of "evolutionary convergence," which Conway Morris defines as "the recurrent tendency of biological organization to arrive at the

---

33. Gould, *Full House*, 214, 215–16.

same 'solution' to a particular 'need.'"[34] Put philosophically, this idea of convergence implies that there are only certain *real* possibilities, in contrast to countless *logical* possibilities, that can actually emerge from the evolutionary process. For Dennett, this narrowing of possible outcomes is due to the fact that evolution proceeds according to algorithmic and genetic processes that lead inevitably from below toward a relatively narrow range of possible outcomes. "The actual genomes that have ever existed," says Dennett, "are a vanishingly small subset of the combinatorially possible genomes." It is this relatively small subset of genomes that defines the range of real, possible evolutionary outcomes. And from this relatively narrow range, the mindless algorithmic recipes of evolution tend to proceed along certain convergent paths in design space. "There is incessant *local* improvement [via evolutionary processes]," Dennett suggests:

> This improvement seeks out the best designs with such great reliability that it can often be predicted by adaptationist reasoning. Replay the tape [of evolution] a thousand times, and the Good Tricks [that lead to improved design] will be found again and again, by one lineage or another. Convergent evolution is not evidence of global progress, but it is overwhelmingly good evidence of the power of processes of natural selection. This is the power of the underlying algorithms, mindless all the way down, but, thanks to the [lifting] cranes it has built along the way, wonderfully capable of discovery, recognition, and wise decision. There is no room, and no need, for [any] skyhooks [or teleology of design].
>
> Can it be that Gould thinks his thesis of radical contingency would refute the core Darwinian idea that evolution is an algorithmic process? That is my tentative conclusion. Algorithms, in the popular imagination, are algorithms *for* producing a particular result. [But that is the mistake of popular conception.] As I said [earlier], evolution can be an algorithm, and evolution can have produced us by algorithmic process, without its being true that evolution is an algorithm *for* producing us.[35]

So for Dennett, Darwin's dangerous idea is that evolution time and again would inevitably produce human beings or something quite similar without the process being designed *for the sake of* producing humankind or anything as such. Thus, according to Dennett, humans find themselves

---

34. Conway Morris, *Life's Solution*, xii.

35. Dennett, *Darwin's Dangerous Idea*, 124, 308. In the initial quotation from 124, Dennett capitalizes "Vanishingly" in the middle of a sentence but, for stylistic reasons, I have used the standard lowercase.

inevitably brought forth or thrown into a mechanistic, mindless, and valueless universe. This is the universal acid of Darwin's dangerous idea. Dennett further speculates that it is "Gould's religious yearnings," in spite of Gould's professed agnosticism, that drive Gould to resist this dangerous idea. That is, given Gould's openness to the possibility of affirming some notion of universal or cosmic value, albeit strictly bifurcated from the realm of science, Gould is resistant to viewing human life as the inevitable product of an evolutionary process of Nietzschean-like eternal recurrence. Instead, Gould prefers to view evolution as a wholly contingent affair, which just happened to produce human beings. Better to be contingent accidents that can hold out the slim hope of some separate realm of meaning than to be eternal prisoners brought forth by a mindless and valueless algorithmic process.[36]

Conway Morris, an evolutionary paleobiologist at Cambridge, shares Dennett's affirmation of evolutionary convergence but does so from an underlying theistic rather than atheistic outlook. One of Conway Morris's central aims is "to argue that, contrary to received wisdom [from Gould and others], the emergence of human intelligence is a near-inevitability." Like Dennett, he sees evolutionary development moving inexorably along certain paths due to convergence, not due to any teleology of design. In accord with this, Conway Morris outlines four overarching conclusions:

> First, what we regard as complex is usually inherent in simpler systems; the real and in part unanswered question in evolution is not novelty *per se*, but how it is that things are put together. Second, the number of evolutionary end-points is limited: by no means everything is possible. Third, what is [really] possible has been arrived at multiple times, meaning that the emergence of the various biological properties is effectively inevitable. Finally, all this takes time. What was impossible billions of years ago becomes increasingly inevitable: evolution has trajectories (trends, if you prefer) and progress is not some noxious by-product of the terminally optimistic, but simply part of our reality.[37]

Like Dennett, Conway Morris thinks Gould's view of radical contingency is mistaken. But instead of drawing acidic metaphysical conclusions from evolutionary convergence, Conway Morris affirms the unity of life and value. Like Darwin, he finds the evolutionary process and its results "genuinely

---

36. Ibid., 309, 311.
37. Conway Morris, *Life's Solution*, xii–xiii.

awe-inspiring. Could it be," he asks, "that attempts to reinstall or reinject notions of awe and wonder are not simply delusions of some deracinated super-ape, but rather reopen the portals to our finding a metaphysic for evolution? And this in turn might at last allow a [genuine] conversation with religious sensibilities rather than the more characteristic response of either howling abuse or lofty condescension."[38]

My contention is that this metaphysic for evolution is to be found by making explicit and further developing the implicit teleology of value in Darwin's own writings. Toward this end, I judge that I am neither qualified on the biological front nor required on the philosophical side to take an unequivocal stance in this contemporary debate.[39] Nonetheless, let me offer at least some constructive observations as a way forward from this important conversation. On the one hand, I am persuaded by Gould's argument for contingency insofar as it makes clear that the evolution of human life is neither necessary nor inevitable from the perspective of a teleology of value. In contrast to a teleology of design, which suggests that nature comes with prefabricated ends, a teleology of value insists that the evolutionary process does not have or require any preordained outcomes. The world could have evolved quite differently and the underlying metaphysic of value would still be the same (I will develop this in chapter 5). Nonetheless, this does not deny that we as human beings are in fact valuing creatures that seek out worth, beauty, and value in various ways, and, just as important, that there is evidence of this seeking out in some other species, albeit to a lesser degree. Thus, on the other hand, what is most intriguing and persuasive about the case for evolutionary convergence is the fact, as Conway Morris notes, that what we see in the complexity of some species is usually latent or inherent in simpler organisms or systems. This point is articulated at great lengths by geneticist Sean Carroll in his fascinating study of evolutionary developmental biology (Evo Devo). "The surprising message from Evo Devo," Carroll announces, "is that all of the genes for building large, complex animal bodies long predated the appearance of those bodies in the Cambrian Explosion. The genetic potential was in place for at least 50 million years, and probably a fair bit longer, before

---

38. Ibid., 5.

39. Physicist Paul Davies takes up this debate between the contingency and inevitability of life on an astronomical level, i.e., whether life on earth is a contingent fluke or one of many inevitable manifestations of life in the cosmos. See Davies, "Are We Alone in the Universe?"

large, complex forms emerged."[40] This reinforces Darwin's central point about the interconnectedness of all life via our common descent. What we thought until quite recently was "junk DNA" appears in fact to contain the vital "switches" that engage or disengage specific combinations of genetic potential.[41] Hence, what is explicitly developed in human beings is latent or "switched off" in the genetic potential of other species. "The shared genetic tool kit for development," notes Carroll, "reveals deep connections between animal groups that were not at all appreciated [before] from their dramatically different morphologies." But this common genetic tool kit does not mean, contrary to Dawkins, that genes drive evolution. "It is clear," Carroll adds, "that genes per se were not 'drivers' of evolution. The genetic tool kit represents possibility—realization of its potential is ecologically driven."[42] Hence, it appears that one can affirm contingency in terms of what potentiality in fact gets actualized and, at the same time, affirm that what is explicit in human behavior is not some outlier in nature but rather is the manifestation of a potential latent in other species. Thus domestic selection, as an expression of our pursuit of value, is indeed a genuine part of nature and not some artificial or superficial epiphenomenon.

SUMMARY

In this chapter I have sought to show the central role that domestic selection plays in Darwin's thinking, that domestic selection is a teleogical activity illustrative of humans pursuing value, and that such activity is neither merely a veiled form of survival or reproductive strategies nor an epiphenomenal excess superfluous to or outside of nature. Rather, the pursuit of value, as explicitly evidenced in human behavior, is the manifestation of a potentiality inchoate or latent in other aspects of the evolutionary process. In the next chapter, I will focus on Darwin's discussion of the value and pursuit of beauty by humans and some species of birds, and its relation to his understanding of sexual selection.

---

40. Carroll, *Endless Forms*, 139.
41. Kolata, "Bits of Mystery DNA."
42. Carroll, *Endless Forms*, 285, 286.

3

BEAUTY AND THE APPRECIATION OF VALUE
The Continuity of Life and Sexual Selection

One of Darwin's lasting insights was his recognition of continuity between humans and other species. Instead of viewing humans as set apart by divine creation, Darwin discerned that we evolved from a common tree of life. Hence, the difference between other species and us is fundamentally a matter of degree, not kind. The degree of difference may be significant or not, but it is never absolute. This premise lies at the heart of Darwin's later major work, *The Descent of Man*, which he published in 1871, just over a decade after *Origin*. If *Origin* was by his own reckoning an abstract, then *Descent* is a more full-bodied account, one that gives a detailed analysis of the place of humans and other species within the overall evolutionary scheme. "The main conclusion" of this work, declares Darwin, "is that man is descended from some less highly organized form. The grounds upon which this conclusion rests will never be shaken, for the close similarity between man and the lower animals ... [is grounded in] facts which cannot be disputed." This commonality is witnessed, he argues, by their homologous embryonic development, by the presence of rudimentary (useless) and nascent (partially developed) organs in humans and other species, and by their many similar behaviors. It is this behavioral similarity that I want to focus on here.[1]

---

1. Darwin, *Descent of Man*, 676. For Darwin's discussion of their homologous embryonic development and the presence of rudimentary and nascent organs, see *Descent*, chap. 1.

Beauty and the Appreciation of Value
INTERPRETING THE CONTINUITY OF LIFE

If one follows Darwin in recognizing continuity between *(a) other species* and *(b) humans*, then the critical question becomes: how does one interpret this continuity? Does one principally interpret *(a) in terms of (b)*, or does one mainly interpret *(b) in terms of (a)*? That is, does one primarily interpret other species as showing behaviors partly similar to or approximating human behaviors, or does one first and foremost interpret human behaviors in terms of the behaviors of other species? The danger of the former is anthropomorphism while the danger of the latter is truncation and reductionism. Yet, if one is going to maintain continuity, as Darwin rightly argued, then one has to understand and interpret this continuity in some manner. What is at stake here is nothing less than the difference between a three-dimensional and a two-dimensional reading of Darwin. A two-dimensional reading opts for interpreting *(b)* strictly in terms of *(a)*—interpreting human behavior as really nothing but what is going on in other species. Hence, from a two-dimensional perspective, human pursuits of beauty, value, and meaning are in fact really nothing but either elaborate strategies for survival and reproduction, or merely epiphenomenal excess that reveals nothing about the character of nature or reality as such. To be sure, there are times when Darwin follows this path and interprets human behavior in terms of other species. This is perhaps best illustrated by his interpretation of morality as a social instinct that is geared toward pleasing others and fostering group loyalty and cohesion. "As man is a social animal," Darwin describes, "it is almost certain that he would inherit a tendency to be faithful to his comrades, and obedient to the leader of his tribe; for these qualities are common to most social animals." In brief, Darwin describes his analysis of morality "as an attempt to see how far the study of the lower animals throws light on one of the highest psychical faculties of man."[2]

But on numerous other occasions Darwin bucks this strict two-dimensional account and goes the other direction and interprets *(a)* in terms of *(b)*: when he discerns in other species behaviors and emotions similar to what is witnessed more explicitly in human experience. It is this three-dimensional tendency in Darwin, I submit, that is most consistent with his overall argument, starting with domestic selection, and with the full panoply of evidence, human and otherwise. The intuition here, as evidenced in the previous chapter by Carroll's discussion of recent findings

---

2. Ibid., 132–33, 120.

in evolutionary biology, is that what is more explicitly visible in one species, such as humans, may be present in a latent, inchoate, or rudimentary form in other species. Whatever differences Darwin may have had with the German naturalist Karl Ernst von Baer (1792–1876), he agreed with him on this point: "Von Baer taught us, half a century ago," observes Darwin, "that, in the course of their development, allied animals put on, at first, the characters of the greater groups to which they belong, and, by degrees, assume those which restrict them within the limits of their family, genus, and species."[3] For Darwin, this certainly does not imply any teleology of design; rather it simply indicates that continuities may be visible in differing degrees of exemplification among differing species. It allows *(b)* to express a distinct or greater capacity without truncating it back into *(a)*. In terms of contemporary evolutionary-developmental biology, this may mean that some combination of genetic switches are turned on for *(b)*, which allow it to exhibit to a greater degree certain characteristics than are exemplified in *(a)*. With this said, let us look at multiple examples of Darwin interpreting *(a)* in terms of *(b)*—of interpreting behaviors in other species as similar or analogous to what is witnessed more explicitly in human experience.

"The lower animals, like man," he states, "manifestly feel pleasure and pain, happiness and misery. Happiness is never better exhibited than by young animals, such as puppies, kittens, lambs, etc. when playing together, like our own children." Here Darwin is clearly interpreting the behavior of other species in relation to human experience, seeing traces of emotional sentiment in these other creatures. He later continues: "Most of the more complex emotions are common to the higher animals and ourselves . . . This shews that animals not only love, but have desire to be loved. Animals manifestly feel emulation. They love approbation or praise; and a dog," he adds, is capable of showing "pride, . . . , modesty, . . . magnanimity, . . . [and] a sense of humour, as distinct from mere play."[4] In *Origin*, Darwin illustrates this continuity between other species and humans in terms of the capacity of reason when he approvingly cites the French naturalist M. Pierre Huber (1777–1840): "A little dose, as Pierre Huber expresses it, of judgment or reason, often comes into play, even in animals very low in the scale of nature."[5] Back in *Descent*, Darwin further articulates this common-

---

3. Ibid., 238.

4. Ibid., 89, 92.

5. Darwin, *Origin*, 234. Recent research appears to confirm Darwin's point about the cognitive capacities of other species, including reptiles: see Anthes, "Coldblooded Does

ality of reason: "Of all the faculties of the human mind, it will, I presume, be admitted that *Reason* stands at the summit. Only a few persons now dispute that animals possess some power of reasoning . . . It is a significant fact," he adds, "that the more the habits of any particular animal are studied by a naturalist, the more he attributes to reason and the less to unlearnt instincts." Darwin extends this capacity further by turning his attention to the possibility of progressive improvement. "It has been asserted [by some,]" he remarks, "that man alone is capable of progressive improvement; that he alone makes use of tools or fire, domesticates other animals, or possesses property; that no animal has the power of abstraction, or of forming general concepts, is self-conscious and comprehends itself." But Darwin retorts: "To maintain, independently of any direct evidence, that no animal during the course of ages has progressed in intellect or other mental faculties, is to beg the question of the evolution of species." He sums up these common threads of emotion and reason when he writes:

> It has, I think, now been shewn that man and the higher animals, especially the Primates, have some few instincts in common. All have the same senses, intuitions, and sensations—similar passions, affections, and emotions, even the more complex ones, such as jealousy, suspicion, emulation, gratitude, and magnanimity; they practice deceit and are revengeful; they are sometimes susceptible to ridicule, and even have a sense of humour; they feel wonder and curiosity; they possess the same faculties of imitation, attention, deliberation, choice, memory, imagination, the association of ideas, and reason, *though in very different degrees.*[6]

It is evident from these numerous passages that Darwin often interprets *(a)* in terms of *(b)*—he interprets other species in relation to humans by noting both the threads of continuity as well as the differing degrees of capacity or exemplification. "There can be no doubt," he concludes, "that the difference between the mind of the lowest man and that of the highest animal is immense." At the same time, he equally insists, "Nevertheless the difference in mind between man and the higher animals, great as it is, certainly is one of degree and not of kind. We have seen that the senses and intuitions, the various emotions and faculties, such as love, memory,

---

Not Mean Stupid."

6. Darwin, *Descent*, 96, 100–102 (italics added). Darwin's point about the rational capacity of other species is evidenced by recent studies of crows, which show their sophisticated ability to make and use tools in order to get other tools in order to later obtain food. See Fleming, dir., "A Murder of Crows."

attention, curiosity, imitation, reason, etc., of which man boasts, may be found in an incipient, or even sometimes in a well-developed condition, in the lower animals."[7] Most contemporary neo-Darwinists tend to dismiss Darwin's observations here as at best merely quaint, precisely because they insist on a strict two-dimensional view, which interprets human behavior in terms of other species—(*b*) in terms of (*a*)—but not other species in terms of human behavior—(*a*) in terms of (*b*).[8] One may perhaps quibble with Darwin for occasionally going overboard and sliding into anthropomorphic excess, but the important underlying point here is that this interpretive stance—of understanding (*a*) in underlying continuity with and in differing degrees of exemplification of (*b*)—enables Darwin to affirm continuity among species without truncating the evidence of his or our lived experience as thinking, feeling, and valuing human beings. When we see behaviors in other species that resemble our own, the answer is not to deny or reduce the evidence of our own experience but rather to affirm some approximate feeling in other creatures. In fact, as Darwin himself suggests above, this outlook is what is required if one is to take seriously the evolution of species. One certainly does not need a teleology of design in order to recognize differing degrees of exemplification, but one does need to recognize differing degrees of exemplification in order to avoid truncating human experiences of worth, meaning, and value.

This joint affirmation of continuity and differing degrees of exemplification is what the German philosopher Hans Jonas describes as "the principle of qualitative continuity" implied by Darwin's understanding of evolution. "If man was the relative of animals," observes Jonas, "then animals were the relatives of man and in degrees bearers of that inwardness of which man, the most advanced of their kin, is conscious in himself." For instance, Darwin speaks above of animals loving and seeking love rather than truncating our experience of love into some mere survival or reproductive instinct. Likewise, he speaks of animals having some modicum of choice and reasoning capacity, albeit to a much lesser degree than what we experience. The clear tendency in all the above examples is to acknowledge and

---

7. Darwin, *Descent*, 150, 151.

8. For example, in his discussion of Darwin's view of aesthetics, which we will turn to shortly, Coyne describes Darwin's view as "quaint" (Coyne, *Why Evolution is True*, 161). But his description applies equally well here to his assessment of Darwin's observations of the continuity of behavior because Coyne and most neo-Darwinists predominantly interpret human activity in terms of other species (*b* in terms of *a*) rather than other species in relation to human behavior (*a* in terms of *b*).

Beauty and the Appreciation of Value

affirm the capacity of other species rather than to truncate human experience in some mechanistic or materialistic mode. The powerful implication of this, Jonas continues, is that "if inwardness is coextensive with life, [then] a purely mechanistic account of life, i.e., one in outward terms alone, cannot be sufficient."[9] What Jonas means by inwardness points in part to what I mean by a teleology of value, namely, that there is evidence in humans and some other species of a pursuit of value above and beyond the basic needs of survival and reproduction. This pursuit of value becomes even more clearly visible in Darwin's many discussions of the ability of various species, especially among some birds, to seek and appreciate beauty.

THE APPRECIATION OF AND DESIRE FOR BEAUTY

When it comes to a "sense of the beautiful," which Darwin defines as "the pleasure given by certain colours, forms, and sounds," we again see him affirming continuity among humans and other species while, at the same time, noting differing degrees of exemplification. Because humans have a greater capacity for complex thought and association, observes Darwin, they are able to appreciate more complex and integrated forms of beauty. "Obviously no animal would be capable of admiring such scenes as the heavens at night, a beautiful landscape, or refined music; but such high tastes," he reflects, "are acquired through culture, and depend on complex associations." But just because other species cannot enjoy beauty to this same degree or in this same complex manner, does not mean that they lack aesthetic sensibilities or desires altogether. On the contrary, he contends, "when we behold a male bird elaborately displaying his graceful plumes or splendid colours before the female, whilst other birds, not thus decorated, make no such display, it is impossible to doubt that she admires the beauty of her male partner." This "taste for the beautiful" is by no means limited to humans. "With the great majority of animals," no doubt, "the taste for the beautiful is confined, as far as we can judge, to the attractions of the opposite sex." I will discuss sexual selection later in this chapter. But the point I want to make first is that Darwin equally emphasizes that some species also pursue beauty for its own sake. Drawing on the observations of the British ornithologist John Gould (1804–1881), he notes that "the nests of humming-birds, and the playing passages of bower-birds are tastefully ornamented with gaily-coloured objects; and this shews that they must

---

9. Jonas, *Phenomenon of Life*, 57, 58.

receive some kind of pleasure from the sight of such things."[10] Darwin later elaborates on this point at length:

> Mr. Gould states that certain humming-birds decorate the outside of their nests 'with the utmost taste; they instinctively fasten thereon beautiful pieces of flat lichen . . . Now and then a pretty feather is intertwined or fastened to the outer sides, the stem being always so placed, that the feather stands out beyond the surface.' The best evidence, however, of a taste for the beautiful is afforded by the three genera of Australian bower-birds . . . The Satin bower-bird collects gaily-coloured articles, such as the blue tail-feathers of parakeets, bleached bones and shells, which it sticks between the twigs, or arranges at the entrance. Mr. Gould found in one bower a neatly-worked stone tomahawk and a slip of blue cotton, evidently procured from a native encampment. These objects are continually re-arranged, and carried about by the birds whilst at play. The bower of the Spotted bower-bird 'is beautifully lined with tall grasses, so disposed that the heads nearly meet, and the decorations are very profuse.' Round stones are used to keep the grass-stems in their proper places, and to make divergent paths leading to the bower. The stones and shells are often brought from a great distance. The Regent bird, as described by Mr. Ramsey, ornaments its short bower with bleached land-shells belonging to five or six species, and with 'berries of various colours, blue, red, and black, which give it when fresh, a very pretty appearance. Besides these there were several newly-picked leaves and young shoots of a pinkish colour, the whole shewing a decided taste for the beautiful.' Well may Mr. Gould [therefore] say, that 'these highly decorated halls of assembly must be regarded as the most wonderful instances of bird-architecture yet discovered.'[11]

It is obvious here that Darwin makes a concerted effort to emphasize that these birds have real aesthetic desire, appreciation, and enjoyment; in short, they value and seek beauty, as illustrated by the elaborate design and accoutrements of their bowers or nests. When Darwin uses the elegant phrase "the taste for the beautiful," as he often does, he implies that birds or animals have both an appreciation of and desire for beauty, such as a desire for "a strong contrast in colour."[12]

10. Darwin, *Descent*, 114, 116, 115.
11. Ibid., 464–65.
12. Ibid., 554.

Indeed, he also suggests "there is reason to suspect that *they love novelty, for its own sake.*"[13] Again, discerning a thread of continuity between human desires for new fashion and taste and what is witnessed in some other species, Darwin reflects: "We may admit that taste is fluctuating, but it is not quite arbitrary. It depends much on habit, as we see in mankind; and we may infer that this would hold good with birds and other animals." Recognizing how breeders use domestic selection as a means to foster and procure desired qualities, he notes how their aesthetic tastes gradually evolve: "they earnestly desire slight changes, which are considered as improvements, but any great or sudden change is looked at as the greatest blemish."[14] Likewise, the same is true for birds:

> With birds in a state of nature we have no reason to suppose that they would admire an entirely new style of coloration, even if great and sudden variations often occurred, which is far from being the case . . . But this dislike of a sudden change would not preclude their appreciating slight changes, any more than it does in the case of man. Hence with respect to taste, which depends on many elements, but partly on habit and partly on a love of novelty, there seems no improbability in animals admiring for a very long period the same general style of ornamentation or other attractions, and yet appreciating slight changes in colours, form, or sound.[15]

Darwin illustrates this desire for novelty, especially among female birds. "It would appear that mere novelty, or slight changes for the sake of change, have sometimes acted on female birds as a charm, like changes of fashion with us." The most vivid example of this desire for novelty, he suggests, is provided by some species of herons, which Darwin speculates evolved in color due partly to an aesthetic desire for novelty. "Some members of the heron family," he writes,

> offer a still more curious case of novelty in colouring having, as it appears, been appreciated for the sake of novelty . . . It appears therefore that, during a long line of descent, the adult progenitors of the *Ardea asha*, the Buphus, and of some allies, have undergone the following changes of colour: first, a dark shade; secondly, pure white; and thirdly, owing to another change of fashion (if I may so express myself), their present slaty, reddish, or golden-buff tints.

13. Ibid., 116 (italics added).
14. Ibid., 556.
15. Ibid.

> *These successive changes are intelligible only on the principle of novelty having been admired by birds for its own sake.*[16]

This taste for the novel and for the beautiful pertains not only to the visual but also to the auditory dimensions of animal life. "With birds," Darwin remarks, "the voice serves to express various emotions, such as distress, fear, anger, triumph, or mere happiness." This expression of happiness comes through song. Again, to be sure, "the true song . . . of most birds and various strange cries are chiefly uttered during the breeding season, and serve as a charm, or merely as a call-note, to the other sex." But Darwin insists that there is "nothing more common than for animals to take pleasure in practicing whatever instinct they follow at other times for some real good."[17] Two things are apparent here: first, Darwin suggests that both males and females can enjoy and appreciate the aesthetic aspects of birdsong (the male in practicing and singing and the female in hearing); and, second, one sees an example here of Darwin slipping from a three-dimensional into a two-dimensional mind-set when he implies that singing for the sake of wooing a female is a "real good" whereas the value and enjoyment of singing for practice is merely, by contrast, an apparent good. But given his unequivocal affirmations above of some birds loving aesthetic novelty and beauty for their own sake, Darwin ought to have stated his point more precisely in terms of differing degrees of value, noting perhaps that singing during mating season is a greater good whereas singing for practice in other seasons is a real, albeit lesser good. Such a formulation would have made the same point with greater overall coherence and consistency. Indeed, a few lines later Darwin proceeds back on course when he concludes that "it is not at all surprising that male birds should continue singing for their own amusement [i.e., for the value of their own enjoyment,] after the season of courtship is over."[18] In emphasizing this capacity of birds to appreciate the beauty of song, he also notes:

> There can be no doubt that birds closely attend to each other's song. Mr. Weir has told me of the case of a bullfinch which had been taught to pipe a German waltz, and who was so good a performer that he cost ten guineas; when this bird was first introduced into a room where other birds were kept and he began to sing, all the others, consisting of about twenty linnets and canaries, ranged

---

16. Ibid., 554, 555 (italics added in the last line).
17. Ibid., 417, 419.
18. Ibid., 419.

themselves on the nearest side of their cages, and listened with the greatest interest to the new performer.[19]

What is particularly noteworthy here is that the aesthetic admirers (linnets and canaries) are of a different species than the singer (bullfinch). In fact, Darwin elsewhere remarks that there are "many recorded instances of tamed or domestic birds, belonging to different species, which have become absolutely fascinated with each other, although living with their own species."[20]

The philosopher Charles Hartshorne, who studied ornithology as a dedicated amateur ornithologist, further explores this aesthetic capacity of birds to appreciate and enjoy song in his book *Born to Sing: An Interpretation and World Survey of Bird Song*. Along the lines of Darwin, Hartshorne contends that "birds are by far the closest of all creatures to man in their interest in sound patterns and skill in their production."[21] Also, in accord with a three-dimensional reading of Darwin, Hartshorne draws continuity between other species and humans by interpreting the former in relation to the latter, rather than the other way around. "With most philosophers and probably most scientists," remarks Hartshorne, "I find strict behaviorism inadequate, at least in the study of human beings; moreover, in view of the evolutionary continuity of life, and the ideal of a unitary explanation of nature as a whole, it seems an unsatisfactory dualism to make man a mere exception. Hence I deny the final adequacy of a purely behavioral view." Drawing on the philosophy of Charles Peirce, Hartshorne avers that it is a mistake to think that aesthetic ideas are merely phenomena restricted to human culture rather than underlying principles of nature or reality as such. The key, Hartshorne learns from Peirce, is to properly "generalize the concepts employed in analyzing aesthetic phenomena, concepts like 'harmony,' 'unity in variety,' and 'feeling.' Taken in their full generality, these concepts do not connote [mere human] complications, but can have instances as simple as life itself." The lesson to be drawn from this, Hartshorne continues, is that

> we shall never understand ourselves or any other animal in a fully satisfactory way until we see that all activity is motivated by the sense of possible harmonies and by the flight from the twin evils of discord and monotony. The "curiosity" which animals display is

19. Ibid., 417–18.
20. Ibid., 466.
21. Hartshorne, *Born to Sing*, 32 (italics removed from original).

relevant here. Novelty is not boring, but it may be either disturbing or pleasing, depending partly upon broadly aesthetic factors of contrast and unity, including contrast and unity between present and in some fashion remembered past experiences.[22]

Hartshorne seeks to further define this underlying aesthetic in terms of two aspects or axes: one involves the polarity between simple and complex while the other involves the range between mere unity and sheer diversity. "In both dimensions," he contends, "beauty [or aesthetic value] is the mean between extremes." Beauty involves some combination of contrast and novelty in the midst of some order and repetition. Hartshorne proceeds to use this evaluative scheme to analyze and interpret different forms of birdsong. Of course, like Darwin, he recognizes that "specifically human aesthetic responses are beyond the capacity of the other animals ... not because the responses are aesthetic, but [rather] because of the intellectual element which pervades them." Thus, there is no single or "absolute aesthetic value in any physical phenomenon, such as a sequence of sounds, which is the same for all types of observers. What is complex or profound for a bird may be ultra-simple for a man. What is ordered for a man may be a meaningless chaos for the bird." Again, in keeping with Darwin's three-dimensional approach and Jonas's emphasis on qualitative continuity, Hartshorne maintains that "higher types of organisms [like humans] can have some sympathy and understanding for lower types. Were this not so there would be no science of behavior! We human beings can appreciate all the degrees of complexity in animal songs, provided our ears (if necessary aided by slowed-down replay) can perceive this complexity. It is the reverse relation which is not possible," i.e., birds cannot understand or appreciate the full complexity of human music.[23] This is simply to reiterate Darwin's point that one should not expect other species to appreciate or experience beauty in the same complex way or to the same degree as humans are capable.

"The basic problem in music and all aesthetic creation," remarks Hartshorne, "is to combine repetition with novelty, the expected with the unanticipated." At the most basic level of "trills or monotones," the only unanticipated features "are when the song is to begin and the number of reiterations. This is a very rudimentary form of novelty." Hartshorne notes, however, "Most birds do better than that." He proceeds to categorize birdsong patterns into six distinct types: (1) "Simple repetition," (2) "Repetition

---

22. Ibid., 1, 2.
23. Ibid., 6, 2, 7–8.

with variations," (3) a "Repertoire of repeated songs," (4) a "Repertoire plus variations," (5) a "Variable sequence," and (6) a "Medley-sequence." With the possible exception of the first category, he suggests that the most "outstanding or 'best' songsters of the world are divided among" these types. To be sure, each of these types has some aesthetic value in terms of order, contrast, and/or novelty; thus, "It seems idle to argue as to which type is better or shows higher development in singing skill. Again, how can one prove that purity of tone and exquisiteness of harmonic contrasts do, or do not, make up for lack of variety?" Just as with human music, "We must not demand all the merits in high degree in each work of art, or we shall never be content with any."[24] Nonetheless, one should not lose sight of Hartshorne's underlying emphasis, which is that some or many birds do sing in a manner that exhibits aesthetic qualities and value and thus one can, as both he and Darwin do, reasonably infer that at least some birds seek and enjoy, to some degree, the aesthetic dimensions of song. This aesthetic element is partly suggested by the tendency among birds to avoid sheer and endless repetition in their songs, which Hartshorne calls "the anti-monotony principle." As he puts it, "The point of the anti-monotony principle is not that birds cannot sing mechanistically, merely repetitively, somewhat as they can fly, and they and we can walk. The point is that they can, and rather generally do, sing nonmechanically, more or less aesthetically," by which he means that birds have some degree of "sensitivity to the value of contrast and unexpectedness as balancing the value of sameness and repetition."[25] Like Darwin, Hartshorne sees no conflict between birds singing for pleasure and singing to maintain territory or to attract mates. These latter two activities, which are certainly important, do not exclude the former. As Hartshorne remarks, "The more essential an activity in the whole life of the bird, the greater the portion of the bird's pleasure which is realized in that activity." Hartshorne concludes the book by suggesting that it is a serious mistake to overlook the desire for and appreciation of aesthetic value in nature:

> Harmonious, intense (sufficiently varied, not too regular) experience is what animals, including human beings, like when they have it and miss when they do not. Not to see this is to overlook much in one's view of life, including subhuman life of all kinds. Aesthetic blindness is more than a superficial defect. It makes all of our science less illuminating than it might be ... Basically what

---

24. Ibid., 90, 91, 92.
25. Ibid., 135, 136.

are good are good experiences, harmonious and intense. Artists, more than others, directly aim to create such experiences, for themselves and others. It is a stupendous fact about nature that the territorial disputes of thousands of species are something like artistic contests—song duels. The struggle is mainly musical (counter-singing), not pugilistic.[26]

There is nothing here, I judge, that Darwin would disagree with; in fact, I suspect that he would agree with all of it. With further empirical evidence and with greater philosophical precision, Hartshorne amplifies Darwin's central point—that there is a real aesthetic desire for and appreciation of beauty among some species besides humans. In short, there is a genuine *taste for the beautiful* exemplified in nature. Again, to be sure, both Hartshorne and Darwin recognize that for most birds and nonhuman species this appreciation of beauty is limited in scope to their attraction to the opposite sex, which leads us now to a discussion of sexual selection.

SEXUAL SELECTION

When he first introduces the concept of sexual selection in *Origin*, Darwin notes that it involves not a struggle for existence—i.e., it is not about characteristics most adaptive for survival—but rather involves "a struggle between the males" for the opportunity to breed with the females. He observes that this rivalry may be defined in terms of the general strength and vitality of the males, or based on their "special weapons" or armaments for battle, or based on their respective beauty (plumage) or aesthetic capabilities (singing capacity). It is this third type of competition that I want to attend to here. "Amongst birds," Darwin writes,

> the contest is often of a more peaceful character. All those who have attended to the subject, believe that there is the severest rivalry between the males of many species *to attract by singing* the females. The rock-thrush of Guiana, birds of Paradise, and some others, congregate; and successive males display *their gorgeous plumage* and perform strange antics before the females, which standing by as spectators, at last *choose the most attractive* partner.[27]

---

26. Ibid., 153, 154, 227.
27. Darwin, *Origin*, 136, 137 (italics added).

In this passage, Darwin is clearly indicating that the female birds select males based on the female's aesthetic desires and judgments and, in knowing this, the males seek to attract the females by fully displaying their beauty. In *Descent*, where he discusses sexual selection in greater detail, Darwin clarifies and amplifies these points. In terms of clarification, he states that when we speak of females choosing males, we are speaking "only from analogy." As he puts it, "It is not probable that [the female bird] consciously deliberates; but she is most excited or attracted by the most beautiful, or melodious, or gallant males." Here, once again, Darwin wants to emphasize that the difference between humans and birds (and other species) in terms of mental powers and aesthetic sensibilities is a matter of degree and not fundamentally one of kind. Hence, though birds may not consciously deliberate, they do, like humans, desire, seek, and respond to aesthetic goods. In terms of amplification, Darwin emphasizes that birds "have acute powers of observation, and they seem to have some taste for the beautiful both in colour and sound."[28] He proceeds to offer further examples of this aesthetic appreciation. "All naturalists who have closely attended to the habits of birds," he reports, "are unanimously of [the] opinion that the males take delight in displaying their beauty." For instance, the male Argus pheasant has "immensely developed secondary wing-feathers" that are

> elegantly marked with oblique stripes and rows of spots of a dark colour, like those on the skin of a tiger and leopard combined. These beautiful ornaments are hidden until the male shews himself off before the female. He then erects his tail, and expands his wing-feathers into a great, almost upright, circular fan or shield, which is carried in front of the body. The neck and head are held on one side, so that they are concealed by the fan; but the bird in order to see the female, before whom he is displaying himself, sometimes pushes his head between two of the long wing-feathers.[29]

In contrast to the Argus pheasant and other birds that display their various forms of beauty, Darwin also notes that some species, such as "the dull-coloured Eared and Cheer pheasants," which lack aesthetic qualities, do not parade themselves in such a manner. In fact, he suggests, such "birds seem conscious that they have little beauty to display." Conversely, some male birds, like the peacock, enjoy showing their beauty to others even when no females are around. As Darwin reports, the peacock "evidently wishes for

28. Darwin, *Descent*, 474, 473.
29. Ibid., 444, 448.

a spectator of some kind, and, as I have often seen, will shew off his finery before poultry, or even pigs." On the whole, Darwin points out that whenever "the sexes differ in beauty, or in the power of singing, or in producing what [might be called] instrumental music, it is almost invariably the male who surpasses the female [in such qualities]." And the reason for this, he concludes, is "that it is the object of the male to induce the female to pair with him, and for this purpose he tries to excite or charm her in various [aesthetic] ways."[30]

The evolutionary implications of this pursuit of and response to aesthetic desire, Darwin suggests, is analogous to what occurs through both domestic and unconscious selection. That is, "if man can in a short time give elegant carriage and beauty to his bantams [domestic game-birds], according to his standard of beauty, [then] I can see no good reason," Darwin reflects in *Origin*, "to doubt that female birds, by selecting, during thousands of generations, the most melodious or beautiful males, according to their standard of beauty, might produce a marked effect [on the breed]."[31] Just as breeders intentionally seek to achieve some desired value in the short-term and, by implication, unconsciously change the breed over the long-term, so too, Darwin reasons, do birds and other species that have a taste for the beautiful unconsciously change their breed over the long run through their immediate pursuit of aesthetic desire. As female birds time and again select the males with the most aesthetic value in terms of beauty or song, these capacities and characteristics tend to get reproduced in the male offspring and thus gradually tend to become enhanced in the breed over time. As illustrated by the famous case of the peacock's tail, it is worth repeating that this gradual enhancement is the result of aesthetic selection and is not for the sake of survival. That is, the large and brightly colored tail of the peacock has been enhanced over time because of the females' desire for beauty; thus they select the males with the most beautiful plumage. But, at the same time, this colorful tail makes the male peacock more visible and vulnerable to predators. If it were not for his ability to fold in the tail, the male peacock would not sufficiently survive to pass on his aesthetic charms.

The upshot of these examples from Darwin pertaining to sexual selection is that there is *real* aesthetic desire and appreciation—a *genuine* taste for the beautiful—at work in nature; it is because the female discerns and is attracted to the beauty of the male that she reproduces with him, not the

---

30. Ibid., 450–51, 444, 455.
31. Darwin, *Origin*, 137.

other way around. She copulates with the male because of her attraction to his beauty; she does not deem him beautiful because she copulates with him. This is clearly Darwin's point, and it is certainly the point of a three-dimensional reading of him.

In contrast, the predominant, two-dimensional reading tends to view the aesthetic element here as merely heuristic, as merely a mechanistic marker indicating good reproductive genes. On this account, it is not really about beauty at all; rather, it is about reproduction and passing on genes. Beauty and aesthetics are merely glossy tools in the service of this one true end. Hence, by implication, the same is true of human relations. Beauty is a heuristic marker for reproductive health, not a genuine value in its own right. Likewise, the pursuit of aesthetic value in art, dance, music, architecture, or other cultural expressions is always either a disguised form of reproductive-oriented activity or it is merely epiphenomenal excess. It is this reductive and truncated understanding of nature and human life that is at odds with a full-bodied reading of Darwin and, more important, at odds with the full texture of our lived experience as human beings in the world.

Even the evolutionary biologist Jerry Coyne acknowledges that the evidence for this genetic-dominant view is suspect. As he reports: "A fair number of studies have found *no* association between mate preference and the genetic quality of the offspring. Still, the good-genes model remains the favored explanation of sexual selection. This belief, in the face of relatively sparse evidence, may partly reflect a preference of evolutionists for strict Darwinian explanations—a belief that females must somehow be able to discriminate among the genes of males." When Coyne refers to a "preference" for "strict Darwinian explanations," he implies that such a preference presumes a strict two-dimensional, mechanistic reading of Darwin; that is, ruling out in advance the possibility that the pursuit of beauty and value could be a real part of nature, such explanations assume from the outset that what can only be going on in nature must be tied narrowly to genetic reproduction and survival. For his part, Coyne, who espouses a "sensory-bias model," makes a partial nod toward affirming Darwin's aesthetic observations, but, in the end, he too makes aesthetic aspects of nature merely instrumental in the service of survival and reproduction. As Coyne writes, "natural selection may often create preexisting [aesthetic] preferences that help animals survive and reproduce, and these preferences can be co-opted by sexual selection to create new male traits. Maybe Darwin's theory of animal aesthetics was partly correct, even if he did anthropomorphize female preferences as a 'taste for the beautiful.'"[32] In other words, for Coyne, Darwin's theory of animal aesthetics is partly correct insofar as aesthetics may serve merely a heuristic role as a habitual bias leading birds and other creatures toward behaviors that foster survival and reproduction. But what Coyne appears to rule out is that birds (and by implication seemingly humans as well) could have any genuine taste for the beautiful; according to his lights, there is no real pursuit or appreciation of beauty or value in nature, rather, such apparent aims are merely useful instruments for serving the more narrow ends of survival and reproduction. In sum, Coyne illustrates another, albeit more nuanced, version of a reductionistic, two-dimensional account of nature.

REFLECTIONS ON REDUCTIONISM

Dennett, among others, would push me to elaborate further on my repeated use of the term *reductionism*. By his lights, reductionism properly conceived is a good thing in contrast to what he calls "*greedy reductionism*, which is not." By the former, he appears to mean an overarching conception of nature devoid of any trace of design, teleology, or God. "The desire to reduce, to unite, to explain it all in one big overarching theory," he suggests, is a worthy goal. "Darwin's dangerous idea is reductionism incarnate, promising to unite and explain just about everything in one magnificent vision," which for Dennett is defined in terms of "an *algorithmic* process."[33] Such a reduced and unified vision, he alleges, does not eradicate or

> *explain away* the Minds and Purposes and Meanings that we all hold dear. People fear that once this universal acid has passed through the monuments we cherish, they will cease to exist, dissolved in an unrecognizable and unlovable puddle of scientistic destruction. This cannot be a sound fear; a *proper* reductionistic explanation of these phenomena would leave them still standing but just demystified, unified, placed on more secure foundations.[34]

In response, let me begin by saying that I too share a desire "to unite" our understanding of nature and the full panoply of our lived experience into a comprehensive vision. This was Whitehead's point by insisting that

---

32. Coyne, *Why Evolution is True*, 166, 167.
33. Dennett, *Darwin's Dangerous Idea*, 82.
34. Ibid.

truth is finally one; this is why philosophy and religion cannot abdicate the challenge and responsibility to constructively engage modern science. Likewise, I applaud Dennett's insistence that biology is not destiny, that human activity is not merely locked in the service of reproductive ends. For instance, in response to the claim by E. O. Wilson and Michael Ruse that "'Morality, or more strictly our belief in morality, is merely an adaptation put in place to further our reproductive ends,'" Dennett declares: "Nonsense. Our reproductive ends may have been the ends that kept us in the [biological] running till we could develop culture, and they may still play a powerful—sometimes overpowering—role in our thinking, but that does not license any conclusion at all about our current values."[35] Like Dawkins, Dennett thinks that humans have escaped the grip of replicating genes by creating culture. Yet, both Dawkins and Dennett insist that culture itself generates "memes," which are cultural products (language, art, religion, ethics, science, and so forth) that follow "the same fundamental [replicating] process that developed the bacteria, the mammals, and *Homo sapiens*." So are humans then merely prisoners of memes? Dennett says no. Rather, like Darwin (as we will see in the next chapter), Dennett affirms the human capacity of "transcendence." As Dennett writes: "Persons, according to the meme model we have sketched, are just larger, higher entities, and the policies *they* come to adopt, as a result of interactions between their meme-infested brains, are not at all bound to answer to the interests of their genes alone—or their memes alone. That is our transcendence, our capacity to 'rebel against the tyranny of the selfish replicators,' as Dawkins says, and there is nothing anti-Darwinian or antiscientific about it."[36] Here too I concur with Dennett's affirmation of the human capacity for transcendence. The problem with the universal acid of his reductionism, however, begins to become apparent at this point.

First, it should be clearly noted that Dennett, in affirming transcendence, is not affirming human freedom in any common sense of the term, that is, in any sense that the future is open and that we have genuine alternatives from which to choose. On the contrary, as he makes clear in his book *Freedom Evolves*, Dennett believes, as a mechanistic determinist, that the past wholly determines the present, which is to say that the past is sufficient

---

35. Ibid., 470; Dennett quotes Ruse and Wilson, "The Evolution of Ethics," *New Scientist* 17 (October 1985) 50–52 (see Dennett's bibliography, 545).

36. Dennett, *Darwin's Dangerous Idea*, 144, 471; for Dawkins on memes, see chap. 11 in *Selfish Gene*.

to guarantee the present; thus, one's "future is fixed" in an objective or metaphysical sense. Dennett denies any genuine agency or freedom to how one presently receives or responds to the past: the past conditions are wholly sufficient to define the present and future because the present and future cannot be otherwise than what the past makes them. For instance, when I miss a short putt on the golf course, it makes no sense, Dennett insists, to say that I genuinely could have made that putt: on the contrary, that putt was always destined—by the infinite stream of past causes—to be missed; it could not have been made, even before I hit the putt. But given the infinitely complex web of prior causes, Dennett holds, our epistemic grasp of the deterministic unfolding of the past into the present is at best limited; thus, our future is "subjectively open," which is to say, given our ignorance of how the present and future are in fact determined by the past, we can, via our capacity of transcendence, conceive of differing worlds or causal streams (I can imagine another world in which I make that putt). Therefore, we live and act *as if* our future is indeed objectively open when in fact it is not. As Dennett describes it in one example: "Her choice is doubtless determined in advance, in the same sense that all events have strict causes that have causes in turn; but what immediately determines her choice is the interplay of elements that, even if well known in themselves, make the outcome [epistemically] unpredictable when they interact recursively."[37] Hence, she lives and acts *as if* she has freedom of choice among alternatives. So, in sum, freedom, for Dennett, refers to our capacity to envision other (wholly determined) worlds or causal streams in the midst of our ignorance of the complexity of how the past in fact wholly determines the present and future of our given world. I can imagine other worlds in which I make that putt, but the actual world I live in is a world in which I was forever destined to miss it. Thus, at most, freedom, for Dennett is a human construct; it is a cultural idiom, a way that we construe our lives in a wholly determined world.

Second, having insisted that there is no place for any teleology in nature, including a teleology of value, Dennett and his version of freedom have nowhere to go; for Dennett, freedom is like a band of prisoners who break free but have nowhere to run. Dennett might reply that humans are free to go anywhere. But by what measure, one might ask, does one assess which directions are better or worse, good or bad? If Dennett replies by

---

37. Dennett, *Freedom Evolves*, 93, 212; for his discussion of the example of the missed putt (an example first introduced by John Austin), see 75–76.

saying that good or bad is defined merely in terms of those directions that enhance human survival and reproduction, then he has simply put humans back in the biological chains of replication. Alternatively, if he suggests that human preference is the sole measure of value, then it is arbitrary which direction freedom travels insofar as it is arbitrary which preferences humans hold. Hence, it does not matter whether the "freed prisoner" escapes or not, since any place or direction is as arbitrarily good or indifferent as any other. Dennett has offered a thoroughgoing mechanistic-algorithmic metaphysic of nature—one simply about means and wholly devoid of any discussion of value or ends. His conception of human freedom and transcendence contains no intrinsic evaluative standard; all such standards are merely culturally constructed and, hence, are extrinsic to nature and to reality as such. Indeed, for Dennett, all first principles dealing with such evaluative questions are in fact "quite arbitrary principles." This arbitrariness becomes even more evident in his repeated discussions of what he calls "design space."[38]

According to Dennett, Darwin showed us, via the principle of natural selection, how to understand the process of accumulative design in nature without a designer. "The idea that Design is something that has taken [time and] work to create," says Dennett, "and hence has value at least in the sense that it is something that might be conserved (and then stolen or sold), finds robust expression in economic terms." Dennett's use of the term *value*" is interesting here: *value for whom or for what?* Later, he states: "Darwin's central claim is that when the force of natural selection is imposed on this random meandering [i.e., random genetic drift], in addition to drifting there is lifting. Any motion in Design Space can be measured, but the motion of random drift is, intuitively, merely sideways; *it doesn't get us anywhere important* . . . In the absence of natural selection, the drift is inexorably *downward* in Design space." What does he mean here by "important"—what defines importance, worth, or value? Dennett continues: "These intuitions about getting somewhere important, about design *improvement*, about *rising* in Design Space, are powerful and familiar, but are they reliable?" He proceeds to imply that importance is measured in terms of solving problems: "Design work—lifting [in design space]—can now be characterized as the work of discovering good ways of solving 'problems that arise.'" The problems Dennett has in mind here are those of successfully adapting to an environment. His point is that nature is able to solve

---

38. Dennett, *Darwin's Dangerous Idea*, 506, 125 (chapter 6 focuses on "design space").

such problems in a cumulative fashion without a prefabricated design or designer. Thus he writes:

> So Paley was right in saying not just that Design was a wonderful thing to explain, but also that Design took Intelligence. All he missed—and Darwin provided—was the idea that this Intelligence could be broken down into bits so tiny and stupid that they didn't count as intelligence at all, and then distributed through space and time in a gigantic, connected network of algorithmic process. The work must get done, but which work gets done is largely a matter of chance, since chance helps determine which problems (and subproblems and subsubproblems) get "addressed" by the machinery.[39]

Dennett's reference to machinery here at the end is telling, for it reveals his underlying mechanistic conception of nature. Why must the work "get done"? Not because the work adds value or importance to the world, but rather because the machinery of nature, in the form of algorithmic process, pushes on inexorably without aim, mind, or value toward solving problems of adaptation.

Yet, it is still unclear what value or importance has to do with any of this endless problem solving. Is Dennett implying that a world with greater complexity—further advancement in design space—is better than one without such complexity? Is a world with humans better than one without? But better for whom or for what or in what way? To say better for humans is to beg the question, for Dennett's aim is to provide a comprehensive view of nature, not just a humanistic projection onto it. Dennett begins to respond to such questions as follows: "There is no single summit in Design Space, nor a single staircase or ladder with calibrated steps, so we cannot expect to find a scale for comparing amounts of design work across distant developing branches." Dennett's conclusion here is somewhat surprising. Based on his previous talk about moving further into design space, one would expect him to say that there is no single way to move upward in design space; hence, there is no single summit. This would be akin to Hartshorne's point that there are numerous ways to achieve aesthetic value in music or song; there is no single path but there is an overarching scale for measuring diverse forms of worth. But, instead, Dennett denies that there is any scale in nature for comparing the value of design across distant branches. Thus he continues:

39. Ibid., 72–73, 125 (some italics added), 126, 133.

Beauty and the Appreciation of Value

> Some lineages get trapped in (or are lucky enough to wander into—take your pick) a path in Design Space in which complexity begets complexity, in an arms race of competitive design. Others are fortunate enough (or unfortunate enough—take your pick) to have hit upon a relatively simple solution to life's problems at the outset and, having nailed it a billion years ago, have had nothing much to do in the way of design work ever since. We human beings, complicated creatures that we are tend to appreciate complexity, but that may well be just an aesthetic *preference* that goes with our sort of lineage; other lineages may be as happy as clams with their ration of simplicity.[40]

It is evident from this revealing passage that complexity is merely an arbitrary human preference ("take your pick") and not a measure of value intrinsic to nature or reality as such. That is, unlike Hartshorne and Peirce, who argue that aesthetic qualities are intrinsic aspects of nature or reality itself, Dennett confines aesthetic value merely to the sphere of human culture. And, unlike Darwin, who affirms genuine aesthetic sensibilities across multiple species, Dennett appears to confine such sensibilities merely to the realm of human preference. So, in the end, his grand unified vision turns out to be merely one of humans projecting their own preferences onto an evolutionary process devoid of value or significance. Dennett proceeds to ask whether the Tree of Life is something sacred. He replies: "Yes, says I with Nietzsche. I could not pray to it, but I can stand in affirmation of its magnificence. This world is sacred." He does not explain what the ground or basis of this sacred worth is, because, by his own lights, it is merely an expression of human wish or preference, and thus it is, as the philosopher and novelist Iris Murdoch might say, an illusion or fantasy we tell ourselves. "There is no denying," Dennett concludes, "that Darwin's idea is a universal solvent."[41] Indeed, on Dennett's reading, it is a solvent that dissolves the ground of meaning and worth. This is precisely what I mean by reductionism, namely, any view that dissolves the ground of meaning and value. Differently stated, any view of nature or evolution that reduces the aesthetic and evaluative aspects of existence either into merely cultural construct or into merely mechanistic means of biological reproduction is a truncated account. Dennett claims that his proper reductionism simply "demystifies"

---

40. Ibid., 134–35 (italics added).
41. Ibid., 520, 521. For Murdoch's insightful discussion about human illusions or fantasies, see Murdoch, *Sovereignty of Good*, 66–70.

meaning and value but leaves them fully intact. A close reading of his view, however, clearly reveals otherwise.

To be sure, Dennett might reply that science properly seeks to offer the most simple or parsimonious explanation of the world and, thus, reductionism is indeed required. It is true that one should avoid excess or redundant hypotheses, but it is equally true that one must fully explain or account for all the facts of experience, including those of value, aim, and meaning. As Peirce learned from Galileo, the simpler explanation does not mean reductionistic or merely logically simpler. Rather, it means simpler in the sense of taking into account the full palette of experience and then offering the most adequate and integrative explanation. Or as Whitehead incisively puts it: "The aim of science is to seek the simplest explanations of complex facts. We are apt to fall into the error of thinking that the facts [themselves] are simple because simplicity is the goal of our [scientific method and] quest. The guiding motto in the life of every natural philosopher should be, Seek simplicity and distrust it."[42] Dennett seeks simplicity in his mechanistic and algorithmic explanation of life and value; his error is that he fully trusts it to be adequate for explaining them.

It is one of the great ironies of modernity that, on the one hand, Copernicus, Galileo, and Darwin rightly decentered humanity: we are not the center of nature or the universe; yet, on the other hand, the secularistic tendencies of modern thought keep implicitly or explicitly reaffirming Protagoras's ancient declaration that *man is the measure of all things*. Why? Because secularistic thinkers like Dennett, who describes his own view as "secular humanism," have nowhere else to turn to try to make sense of the evidence of our everyday experience of value.[43] Hence, they keep making humans more important, and, at the same time, they try to make them unimportant. This tension—if not contradiction—indicates, I would suggest, the problem with a view of nature wholly devoid of any conception of a teleology of value and, indeed, devoid of any conception of theism. I will seek to spell this out more fully in chapter 5, but, before getting there, I will examine in chapter 4 Darwin's conception of what makes human existence distinctive.

42. Peirce, *Selected Writings*, 372–73; Whitehead, *Concept of Nature*, 163.

43. Dennett, *Darwin's Dangerous Idea*, 476. Thomas Nagel and Jeffrey Stout should be noted as two contemporary thinkers who seek to offer a secularistic conception of value that is nonanthropocentric. Yet their respective attempts, as I've argued elsewhere, run into their own problems: for Stout, see Meyer, *Metaphysics and the Future of Theology*, 187–206; and for Nagel, see Meyer, "Value and Conceptions of the Whole."

Beauty and the Appreciation of Value

SUMMARY

In this chapter I have sought to outline Darwin's fundamental point about the continuity between humans and other species, and how some other species exemplify, to some degree, a pursuit of value, especially in the form of beauty. This taste for the beautiful, as Darwin aptly calls it, is evidence, I argue, of a teleology of value (not a teleology of design) at work in nature. Attempts to deny or explain away the role of value in nature I judge to be reductionistic. Modern science has rightly revealed that humans are not the center of the universe; but this means then that neither are humans the ultimate measure of value. A teleology of value requires a deeper ground, one connected finally to an adequate form of theism, that is, to one that can coherently integrate value and evolutionary change.

4

HUMAN EXISTENCE

The Capacity for Understanding and Evaluation

> *It may be freely admitted that no [other] animal [besides man] is self-conscious, if by this term it is implied, that he reflects on such points, as whence he comes or whither he will go, or what is life and death, and so forth.*
>
> —Charles Darwin, *The Descent of Man*

This little noticed passage in the works of Darwin is of significant import because it concisely articulates the distinctive character of human existence.[1] Though Darwin indeed stresses, as we have seen, the threads of continuity between humans and other species, he here summarily delineates what a difference in degree looks like when it comes to the possibilities of human life. It is this capacity for self-conscious awareness—this ability to rise above the present, to reflect on the past, and to anticipate the future—that marks distinctively human existence. Darwin suggests that it is our ability to reflect on existential, scientific, and teleological questions, such as "whence" we come and "whither" we go, that distinguishes us from other species. To be sure, even here he proceeds to speculate that an "old dog" may have some limited degree of self-awareness whereas a hard-working wife of a

---

1. Darwin, *Descent*, 105 (italics added).

"degraded Australian savage" may have very little opportunity to develop her capacity for such reflection.[2] Nevertheless, it is this greater potential for self-conscious awareness, what some call a capacity for self-transcendence, that on the whole distinguishes human existence from other forms of life. It is this ability that enables us to seek to understand the past, to appreciate the present, and to anticipate and pursue future possibilities. Indeed, what Darwin rightly points to is the fact that our capacity for understanding is inextricably tied to our capacity for evaluation; that is, the quest for understanding is itself a teleological activity. As the philosopher Franklin Gamwell observes:

> We humans live with understanding, conscious of both ourselves and other things. While we also live within limits determined by the past we inherit or the environment in which we are set, we nonetheless are aware of alternative ends at which we might aim and thus are able in some measure consciously to decide what we will be or become. Thereby, human life is a moral [i.e., teleological and evaluative] enterprise because understanding alternatives for purpose entails decision among them by way of an evaluation.[3]

For instance, to decide to study nature, as Darwin nobly did, is an evaluation of an aim and a life judged worthy of pursuit. The crucial point is that the capacity to understand, to assess worth, and to pursue value is, by Darwin's own lights, explicitly evident in the human species. As Gamwell emphatically notes, "That existence with understanding has emerged in our world within the human species is an empirical fact."[4] In what follows, I will first seek to unpack and illustrate some of the important implications of Darwin's description of distinctively human existence by using Darwin himself as the chief exemplar. I will then conclude by outlining some of the inherent philosophical suppositions of existence with understanding, specifically, the conditions of subjectivity.

THE QUEST FOR UNDERSTANDING

A powerful image comes to mind as one reads and studies Darwin, namely, envisioning him stepping away from the long sweep of evolutionary

---

2. Ibid.
3. Gamwell, *Existence and the Good*, 17.
4. Ibid., 124.

history, a history that brought him into existence, in order to understand and evaluate the nature and significance of the evolutionary process itself. It is this vision of Darwin stepping back from the process in order to understand and assess the process that symbolizes our human capacity for self-transcendence or self-conscious awareness. This is of no little consequence, for it indicates that the capacity, desire, and aim to know and appreciate is something that becomes manifest in and through the evolutionary process itself. Again, as Whitehead remarks, there is a threefold urge made explicit in nature: to live, to live well, and to live better; that is to say, there is a quest for value and understanding and not just a quest for survival and reproduction.[5] Darwin himself gives evidence of this teleological quest in discussing his desire to understand. Looking back on his life, he reflects: "From my early youth I have had the strongest desire to understand or explain whatever I observed,—that is, to group all facts under some general laws." This desire for a life devoted to the aim of understanding is evident early in his career while onboard the *Beagle*. In a letter to his sister, dated June 1833, he writes:

> I trust and believe that the time spent in this voyage, if thrown away for all other respects, will produce its full worth in Natural History; and it appears to me the doing what *little* we can to increase the general stock of knowledge is as respectable an object [or aim] of life as one can in any likelihood pursue. It is more the result of such reflections . . . than much immediate pleasure which now makes me continue the voyage, together with the glorious prospect of the future, when passing the Straits of Magellan, we have in truth the world before us.[6]

Here we see a clear example of Darwin reflecting on and evaluating the worth of continuing a difficult journey onboard ship *for the sake of* a life dedicated to the aim of increasing the general stock of knowledge. It is not for the immediate pleasure, he insists, but rather for the value of the future goal that he pursues this endeavor. Looking back on this voyage late in life, Darwin ruminates: "As far as I can judge of myself, I worked to the utmost during the voyage from the mere pleasure of investigation, and from my strong desire to add a few facts to the great mass of facts in Natural Science. But I was also ambitious to take a fair place among scientific men, – whether more ambitious or less so than most of my fellow-workers, I can form no

---

5. Whitehead, *Function of Reason*, 8.
6. Darwin, *Autobiography*, 63, 164.

opinion." In these later reminiscences, Darwin appears to have tempered the unpleasant hardships of the voyage and focused instead on the joy and pleasure of the inquiry itself, which was surely part of his earlier experience. Along with this emphasis on the qualitative value of the practice, one still sees him defining his endeavor in terms of pursuing aims, specifically, the aims of scientific knowledge and professional advancement. Four years after returning from the *Beagle*, Darwin wrote to a friend in 1840: "I have nothing to wish for excepting stronger health to go on with the subjects to which I have joyfully determined to devote my life."[7] As Darwin's life illustrates, human endeavor, including science itself, is a teleological activity in pursuit of value, such as knowledge, enjoyment, and advancement.

Though this teleological pursuit of knowledge and understanding surely undergirds and inspires all of Darwin's intellectual endeavors, it can perhaps be illustrated most vividly by noting the evaluative and selective methodological choices he made in two of his works: *Origin* and *The Expression of the Emotions in Man and Animals*. As indicated earlier, *Origin* was by Darwin's own account an abstract and, like any abstract, it intentionally leaves out some matters and details. As historian Joe Cain remarks, "In the rush to complete *On the Origin of Species* (1859) Darwin made some strategic decisions . . . Darwin deliberately dropped several key topics. These, he thought, would simply be too distracting." These omissions included the origins of life and human evolution.[8] The relevant point here is that Darwin's own strategic decisions illustrate that human beings are creatures that evaluate alternative possibilities and choose those that they judge to be good in relation to the aims they pursue. For his part, Darwin judged that his intellectual aims would best be served by making *Origin* a more narrowly focused composition with a more streamlined argument.

As in his other works, Darwin's underlying interest in *The Expression of the Emotions in Humans and Other Animals* (1872) is guided by the question of origins—understanding the "whence" of things. More specifically, *Expression* is the attempt to observe and understand the habitual and involuntary expression of emotions in humans and other animals. Inspired in part by watching his own children's expressions, Darwin sought to catalog and analyze a wide range of human emotional responses, especially strong

---

7. Ibid., 34, 170.

8. Cain, "Introduction" to Darwin's *Expression of the Emotions in Man and Animals*, xi, xii.

and intense ones.[9] At first, as he himself describes it, he "hoped to derive much aid [in this endeavor] from the great masters in painting and sculpture, who are such close observers. Accordingly," he continues, "I have looked at photographs and engravings of many well-known works; but, with a few exceptions, have not thus profited. The reason no doubt is, that in works of art, beauty is the chief object; and strongly contracted facial muscles destroy beauty."[10] Darwin here recognizes that different aims have different needs: artists seek beauty and thus seek subtle and harmonious expression; alternatively, he sought to study intense and raw expression, which mars beauty. Hence, he created his own sample by commissioning photographs to be made of humans of diverse ages expressing various intense emotions. As he describes: "I have found photographs made by the instantaneous process the best means for observation, as allowing more deliberation. I have collected twelve [photographs], most of them made *purposely* for me." Here again one sees that the quest for understanding is a teleological endeavor, one that involves evaluation and selection from among alternative possibilities—in this case, from among alternative methodological means. Likewise, the selection of what to study or what to devote attention to is itself an evaluative undertaking, as Darwin implies in the closing lines of *Expression*: "We have also seen that expression in itself, or the language of emotions . . . is certainly of *importance* for the welfare of mankind. To understand, as far as is possible, the source or origin of the various expressions [of emotion] . . . *ought* to possess much *interest* for us."[11] Here he is clearly assessing the worth and significance of one course of action (studying the origins of the expression of emotions) in contrast to its alternative (not to pursue this endeavor). In brief, those neo-Darwinists and others who deny that human life, as part of life more generally, is a teleological and evaluative enterprise have to deny the purposive nature of their own forms of scientific inquiry, which, as we see here, Darwin's own efforts belie.

---

9. Ibid., xxvi.

10. Darwin, *Expression of the Emotions in Man and Animals*, 25; for a related point about art's devotion to the aim of beauty above accuracy or truth, see ibid., 170.

11. Ibid., 139 (italics added), 334 (italics added).

Human Existence

EVALUATION AND THE EXPRESSION OF EMOTIONS

Though the expression of emotions is not primarily a volitional act, it is nonetheless an important indicator of the evaluative dimension of life. That is, emotions are nothing if not an *evaluative* response to an environment, situation, or state of affairs. As Darwin's study in *Expression* illustrates, the response may indeed be habitual, unconscious, or even an involuntary reflexive remnant from an evolutionary past. Nevertheless, emotions are always actions, expressions, or dispositions tinged with evaluative and affective tone; whether it is bliss, sudden terror, or crushing despair, the emotions express an immediate affective appraisal of a situation. In fact, there is perhaps no more visible evidence of the evaluative dimension of life than the expression of emotions.

For his part, Darwin examines emotions ranging from the tragic and sublime to the ridiculous, and from loving affirmation to loathing rejection. "No emotion is stronger," he observes, "than maternal love," but this love may not always be visibly expressed. When love is expressed, it is commonly expressed through gently touching or caressing the beloved. "A strong desire to touch the beloved is commonly felt," says Darwin, "and love is expressed by this means more plainly than by any other. Hence we long to clasp in our arms those whom we tenderly love. We probably owe this desire," he speculates, "to inherited habit, in association with nursing and tending of our children, and with the mutual caresses of lovers." But whatever the origin of our habitual form of expression, the important point here is that such manifestations are *evaluative* articulations of significant worth and value; a tender caress is an expression and affirmation of great worth. Darwin takes this evaluation for granted when he proceeds to examine emotions of loss and suffering. "When a mother suddenly loses her child," he observes,

> sometimes she is frantic with grief, and must be considered to be in an excited state; she walks wildly about, tears her hair or clothes, and wrings her hands . . . [U]nder the sudden loss of a beloved person, one of the first and commonest thoughts which occurs, is that something more might have been done to save the lost one. An excellent observer, in describing the behaviour of a girl at the sudden death of her father, says she "went about the house wringing her hands like a creature demented, saying 'It was her fault'; 'I should never have left him'; 'If I had only sat up with him'" . . .

> As soon as the sufferer is fully conscious that nothing can be done, despair or deep sorrow takes the place of frantic grief.[12]

Of course Darwin's primary interest is in the physiological aspects of such grief, but my concern is to note the obvious but essential point that such grief is the expression of deep and genuine evaluative worth: it is because we love so deeply that we grieve the loss of the valued other. Indeed, at one point sounding like an existentialist philosopher, Darwin remarks: "The expression . . . of grief and anxiety is eminently human."[13]

Reflecting on less tragic but no less poignant loss, Darwin writes: "The vivid recollection of our former home, or of long-past happy days, readily causes the eyes to be suffused with tears; but here, again, the thought naturally occurs that these days will never return. In such cases we may be said to sympathise with ourselves in our present, in comparison with our former, state."[14] Here Darwin nicely captures the temporal character of our lives, and thus the bittersweet nature of existence with understanding: we experience value and its concrete passing; we then presently recollect what was, such as the joy of our former home, and recognize and mourn its loss. And yet, in some sense, the past value is still a datum of present experience: the past is enveloped in the present, albeit in a fragmentary way. Memory is our conscious awareness of this datum, and emotion, such as tears in our eyes, is an expression of our evaluative sense of its worth. Darwin takes all of this quite for granted: humans are creatures who pursue and respond to life through a teleological and evaluative prism.

This evaluative prism is also evident in Darwin's other discussions of positive, negative, and lighthearted emotions. Describing common signs of evaluation, such as shaking of one's head, Darwin declares: "These signs are indeed to a certain extent expressive of our feelings, as we give a vertical nod of approval with a smile to our children, when we approve of their conduct; and shake our heads laterally with a frown, when we disapprove." Speaking of more sharply negative emotions and evaluations, he states: "We have now seen that scorn, disdain, contempt, and disgust are expressed in many different ways, by movements of the features, and by various gestures; and that these are the same throughout the world. They all consist of actions representing the rejection or exclusion of some real object which we dislike or abhor." Likewise, taking stock of laughter and lightheartedness, Darwin

12. Ibid., 80, 197, 81–82.
13. Ibid., 331.
14. Ibid., 199.

adds: "Yet laughter from a ludicrous idea, though involuntary, cannot be called a strictly reflex action. In this case, and in that of laughter from being tickled, the mind must be in a pleasurable condition; a young child, if tickled by a strange man, would scream from fear. The touch must be light, and an idea or event, to be ludicrous, must not be of grave import."[15] As evident in all three of these sets of examples, Darwin takes for granted that humans evaluate their environments or circumstances and respond accordingly; in order for an idea to be received as ludicrous, it must be judged "not to be of grave import." In order for tickling to be a pleasurable condition expressed by laughter, it must be understood to be from a person known or perceived to be trustworthy. Similarly, expressive vertical or lateral head movements are indicative of positive or negative evaluations of given actions or conditions. So too, expressions of scorn and disgust are visceral assessments of a given situation or state of affairs.[16] Again, it is certainly the case that Darwin's primary interest is in the physiological origin and expression of such emotions; nevertheless, he certainly recognizes and takes for granted that the evaluative dimension is the fundamental aspect of all emotions. As Whitehead remarks, "We see at once that the element of value, of being valuable, of having value ... must not be omitted in any account of an event as the most concrete actual something. 'Value' is the word I use for the intrinsic reality of an event. Value is an element which permeates through and through [a properly inclusive] view of nature."[17] This holistic view is implied by Darwin here in *Expression* and, as I have sought to show, in many other notable places in his writings.

VALUE AND THE APPRECIATION OF BEAUTY

Where Darwin becomes most explicit about the value dimension of life is in his numerous evaluative assessments and expressions of beauty. One cannot attentively read his works without noticing how frequently he offers passing comment on the beauty or lack thereof of the various species or environments he has studied. In his autobiography, reflecting back on his experience on the *Beagle*, he muses: "The glories of the vegetation of the

---

15. Ibid., 251, 239, 186.
16. The neuroscientist V. S. Ramachandran likewise recognizes the evaluative dimension of emotions, namely, that emotional expressions involve an evaluative assessment of a situation or environment; see Ramachandran, *Tell-Tale Brain*, 39–40.
17. Whitehead, *Science and the Modern World*, 93.

Tropics rise before my mind at the present time more vividly than anything else; though the sense of sublimity, which the great deserts of Patagonia and the forest-clad mountains of Tierra del Fuego excited in me, has left an indelible impression on my mind." Later, recounting his intellectual work in the period between being onboard the *Beagle* and the writing of *Origin*, he recalls:

> It was evident that such facts as these [gathered from the *Beagle*], as well as many others, could only be explained on the supposition that species gradually become modified; and the subject haunted me. But it was equally evident that neither the action of the surrounding conditions, nor the will of organisms (especially in the case of plants) could account for the innumerable cases in which organisms of every kind are *beautifully* adapted to their habits of life.[18]

In these recollections, we see evidence of Darwin's evaluative and aesthetic judgments fully embedded in the midst of his biological observations. Whether he is assessing the glories of tropical vegetation, the sublimity of South American deserts and mountains, or the beauty of how organisms are fittingly adapted to their way of life, he clearly expresses and illustrates the human capacity to appreciate value and beauty in the world. Of course, it was Darwin's famous evaluative judgment in *Origin*—that there is grandeur in nature when understood in evolutionary terms—that I used to launch this study. Likewise, we saw in *Descent* countless examples of his appreciation of the beauty of birds and other species.

Darwin's son Francis, who put together the autobiographical volume in 1892, vividly recalls his father's aesthetic enjoyment, especially in the garden at home. Frances recounts:

> Though he took no personal share in the management of the garden, he had great delight in the beauty of flowers—for instance, in the mass of Azaleas which generally stood in the drawing-room. I think he sometimes fused together his admiration of the structure of a flower and of its intrinsic beauty; for instance, in the case of the big pendulous pink and white flowers of Diclytra. In the same way he had an affection, half-artistic, half-botanical, for the little blue Lobelia. In admiring flowers, he would often laugh at the dingy high-art colours, and contrast them with the bright tints of nature. I used to like to hear him admire the beauty of a flower; it was a kind of gratitude to the flower itself, and a personal love

18. Darwin, *Autobiography*, 34, 47 (italics added).

for its delicate form and colour. I seem to remember him gently touching a flower he delighted in; it was the same simple admiration a child might have.[19]

His son's account here poignantly depicts Darwin as a man whose conception of nature, rather than an austere and reductive materialism, was shot through with value and aesthetic quality. He "sometimes fused together his admiration of the structure of a flower and of its intrinsic beauty." His approach was "half-artistic, half-botanical." His admiration of beauty "was a kind of gratitude to the flower itself, and a personal love for its delicate form and colour." It is no wonder that Francis continues by observing that his father "could not help personifying natural things" precisely because Darwin, at his best, saw the presence of value in nature and life. This is simply to say again, with Whitehead, that value is an element that permeates a properly inclusive view of nature.[20]

In contrast to his generally consistent lifelong appreciation of natural beauty, it is interesting to note how Darwin later in life reflects on his diminished capacity to appreciate the value of art, literature, and music:

> I have said that in one respect my mind has changed during the last twenty or thirty years. Up to the age of thirty, or beyond it, poetry of many kinds, such as the works of Milton, Gray, Byron, Wordsworth, Coleridge, and Shelley, gave me great pleasure, and even as a schoolboy I took intense delight in Shakespeare, especially in the historical plays. I have also said that formerly pictures gave me considerable, and music very great delight. But now for many years I cannot endure to read a line of poetry; . . . I have also almost lost my taste for pictures or music . . . I retain some taste for fine scenery, but it does not cause me the exquisite delight which it formerly did . . .
>
> This curious and lamentable loss of the higher aesthetic tastes is all the odder, as books on history, biographies, and travels (independently of any scientific facts which they may contain), and essays on all sorts of subjects interest me as much as ever they did. My mind seems to have become a kind of machine for grinding general laws out of large collections of facts, but why this should have caused the atrophy of that part of the brain alone, on which the higher tastes depend, I cannot conceive . . . [I]f I had to live my life again, I would have made a rule to read some poetry and listen to some music at least once every week; for perhaps the parts of

19. Francis Darwin in Darwin, *Autobiography*, 87.
20. Ibid.; Whitehead, *Science and the Modern World*, 93.

my brain now atrophied would thus have been kept active through use. The loss of these tastes is a loss of happiness, and may possibly be injurious to the intellect, and more probably to the moral character, by enfeebling the emotional part of our nature.[21]

This intriguing passage kindles at least four observations. First, it is certainly the case that our interests in and receptivity to specific goods can evolve or diminish over time. Our particular inclinations, desires, and passions may vary in due course. But, second, Darwin's reflections also call to mind the counsel of John Stuart Mill, whom Darwin had read, namely, that one must continually nurture one's appreciation for higher goods and pleasures by sustained exposure to them. Hence, we see Darwin here mildly chastising himself for not having continued the practice of reading poetry and listening to music throughout his life. As he puts it, "if I had to live my life again, I would have made a rule to read some poetry and listen to some music at least once every week." In the spirit of Mill, he even conjectures that the loss of these higher tastes contributes to a "loss of happiness" and to a weakening of moral character.[22] Third, one can never forget that Darwin and his wife Emma suffered the tragic death of three of their ten children, including their eldest daughter Annie, of whom Darwin was particularly fond. Annie died in 1851 at the tender age of ten; a week after her death, Darwin wrote: "We have lost the joy of the household, and the solace of our old age. She must have known how we loved her. Oh, that she could now know how deeply, how tenderly, we do still and shall ever love her dear joyous face! Blessings on her!"[23] As evidenced by these wrenching words, such an existential and emotional blow could surely affect one's outlook on life and, in turn, could dampen one's aesthetic taste and appreciation. But, fourth, in addition to this existential toll, the passage above insightfully suggests that Darwin's methodological and scientific training may have also contributed to his more narrow and increasingly tone-deaf response to arts and literature; indeed, as he puts it, such a loss may even "be injurious to the intellect." The key line above is when he concludes: "My mind seems to have become a kind of machine for grinding general laws out of large collections of facts, but why this should have caused the atrophy of that part of the brain alone, on which the higher tastes depend, I cannot conceive."

21. Darwin, *Autobiography*, 61–62.
22. For a discussion of Darwin having read J. S. Mill, see Allhoff, "Evolutionary Ethics from Darwin to Moore," 90.
23. Darwin, *Autobiography*, 102.

To unpack this, it is instructive to compare Darwin and Whitehead in their respective assessments of Wordsworth.

Whereas Darwin began by admiring Wordsworth's aesthetic affirmation of value in nature but later lost his appreciation of Wordsworth due to his increasingly mechanistic mind-set, Whitehead persistently points to Wordsworth and the other English romantic poets as a necessary and valuable corrective to the mechanistic outlook of modernity. "We are here witnessing [in the English romantics]," observes Whitehead, "a conscious reaction against the whole tone of the eighteenth century. That century approached nature with the abstract analysis of science, whereas Wordsworth opposes to the scientific abstractions his full concrete experience [of nature]." Wordsworth's insight is to recognize not only how science abstracts from the full texture of our experience of nature but how it then forgets that it has done so, which Whitehead sums up in Wordsworth's famous line: "'We murder to dissect.'" In contrast, Wordsworth "always grasps the whole of nature as involved in the tonality of the particular instance."[24] By his own account, Darwin's mind has become a machine for grinding out abstract laws, regularities, or averages, which, by definition, ignore or leave out the aesthetic value or tone that is part of each concrete instance. Notice, for example, that his explanation of why he has lost his appreciation for higher tastes appeals to a physicalist account (i.e., part of his brain has atrophied); he never considers that his underlying methodological assumptions may have steered him too far away from the full facts and texture of concrete experience. This is why Whitehead argues that the chief role of philosophy is to serve as "the critic of abstractions" in the various academic disciplines and areas of culture. First, it must point out and always remind us that science, in its pursuit of laws and patterns of regularity, is speaking in terms of abstractions; and, second, it must then always seek to complete these abstractions "by direct comparison with more concrete intuitions of the universe, and thereby promot[e] the formation of more complete schemes of thought. It is in respect to this comparison that the testimony of great poets is of such importance. Their survival is evidence that they express deep intuitions of mankind penetrating into what is universal in concrete fact." As illustrated by Wordsworth, the English poetic literature of the nineteenth century "is a witness to the discord between the aesthetic intuitions of mankind and the mechanism of science."[25]

24. Whitehead, *Science and the Modern World*, 81, 83.
25. Ibid., 87.

This discussion presents us with another apparent disjuncture in Darwin: on the one hand, as I have repeatedly suggested, it appears that he was able to appreciate the value of beauty in his own direct experience of nature, as attested by his numerous observations of beauty in birds and as noted by his son Francis regarding his appreciation of the beauty of flowers. Yet, on the other hand, he appears to have gradually lost his ability to appreciate verbal, artistic, and musical expressions of beauty, including those depicting nature, as illustrated by the works of Wordsworth and the other romantic poets. Perhaps Darwin's scientific habits of mind, coupled with the existential loss of Annie and his other children, made it harder for him to appreciate indirect, symbolic, or artistic expressions of beauty whereas he was able to continue to experience and appreciate such beauty in his direct encounters with nature. However one might speculatively explain the cause of this split, this divergence is nevertheless suggestive of the two tendencies evident in Darwin's writings, namely, the three-dimensional version, which intuits and appreciates the aesthetic and teleological dimensions of value in nature, and the more commonly perceived two-dimensional version, which is imbued with the mechanistic and reductionistic tendencies of modern thought in its pursuit of grinding out scientific laws abstracted from the data of concrete fact and experience.

EXISTENCE WITH UNDERSTANDING: THE CONDITIONS OF SUBJECTIVITY

I began this chapter by quoting Darwin's account of what makes human life distinctive in comparison to that of other species, which is our capacity for self-conscious awareness. It is this capacity that enables us to ask existential, scientific, and teleological questions about our lives and the world. I then noted, drawing on the work of Gamwell, how existence with understanding is inherently an evaluative and teleological endeavor. I want to conclude this chapter by returning to Gamwell in order to outline in more philosophical detail what this means.

As Darwin indicated in his account of human existence, the capacity for self-conscious awareness is what is most distinctive about human beings; it is what sets us apart from other species. To be sure, the difference is one of degree, not kind, but the difference is significant, as Darwin rightly observes. This capacity for self-conscious awareness is what Gamwell and other philosophers describe as *subjectivity*. To seek to identify the necessary conditions or characteristics of subjectivity is what philosophers, since the time of Kant, have called *transcendental* analysis. Gamwell offers his own transcendental account, which I will briefly outline here.

"Subjective existence," says Gamwell, "is a specific kind of existence," namely, self-conscious existence. It is "not necessary to all existence," which is simply to repeat that it is a characteristic that denotes existence with understanding and thus distinguishes humans from other species.[26] This is not to say that only human beings can exemplify subjectivity but merely to say that humans, based on our empirical observations of the world, are the only species that we know of that live with a developed understanding. Some birds and other species indeed appear to appreciate and pursue the value of beauty to some degree, but humans can and do pursue such value in a self-conscious manner.

"The features defining subjectivity in general," Gamwell contends, "include certain inescapable understandings." To unpack his meaning here, we must first reflect further on the nature of subjectivity. As was just noted, subjective existence does not define all existence; rather, it is a specification or subset of existence as such. This means that the generic characteristics of existence must be defined in terms apart from subjectivity, consciousness, or understanding. Following Whitehead and Hartshorne, Gamwell defines such characteristics in terms of relations to an order of actualities and possibilities. Hence, subjectivity, as a specification of existence, is "not exhausted by understanding. If understanding presupposes but is not presupposed by beings as a whole," remarks Gamwell, "then exemplifications of subjectivity are constituted by prior or nonconscious relations to an order of actualities and possibilities." In other words, existence as such is defined in terms of relations; what subjectivity distinctively adds is the capacity for awareness or understanding of these relations. In Gamwell's words, these relations are "prior in constituting the subject, such that understanding is a consciousness of these relations. Thus, subjects are *distinguished* by understanding; that is, something can be a subject only if it is also nonconsciously related to an order of actualities and possibilities to which the subject belongs, while something can be so related without also being a subject."[27]

From here, Gamwell notes that subjectivity involves not only explicit awareness, what we consciously focus on in the foreground of our attention, but also implicit awareness of a wider background that sets the context

---

26. Gamwell, *Existence and the Good*, 7.
27. Ibid., 7, 51.

of the foreground. "The complexity of human understanding," Gamwell observes, "cannot be appreciated without a distinction between explicit understandings, those in the foreground of attention, and a background of understandings that are implicit, where 'implicit' here means 'contained in the nature of something although not readily apparent.'"[28] In other words, to recognize and affirm that humans exist with self-conscious awareness, as Darwin rightly does, is implicitly to recognize and affirm in a transcendental sense that human understanding also contains wider and dimmer forms of background awareness. This dim background includes relations to past actualities and to future possibilities. In fact, it is the subject's relations to this wider background, to this wider whole, that enframe and enable conscious foreground experience. As Gamwell puts it, "the background is required in order for the foreground to be what it is."[29] The key point here is that these background relations fundamentally inform the existing subject and thus are prior to any explicit understanding of them. Cast in Darwin's terms, the present human subject is always implicitly related to, as a descendent of, the whole of evolutionary history—cosmological as well as biological; if Darwin taught us anything, he taught us that we are implicitly and internally related to the whole evolutionary past, as symbolized by the tree of life. Furthermore, as a creature existing with self-conscious awareness, the present human subject is also related to a realm of possibilities, which define the alternatives for potential aims and courses of action. "Decision with understanding," Gamwell declares, "is decision for a telos, some future possibility, at the realization of which self-expression is aimed." In short, human beings "enjoy the distinctive freedom they do," he adds, "because they are able to think about ends." Thus in the case of the young Darwin, for instance, he had a choice among medicine, ministry, and science as possible future vocations for his life.[30]

But selection among alternative possible aims implies evaluation of those alternatives, and evaluation necessarily implies some conception of the good as always present in the background of understanding. Gamwell describes this as follows:

> Self-understanding also means that subjective activities necessarily affirm some understanding of the good, in terms of which alternative ends are discriminated as better and worse ... Indeed,

---

28. Ibid., 7, 55; Gamwell here quotes *The American College Dictionary*.
29. Ibid., 55.
30. Ibid., 97; for Darwin's vocational alternatives, see his *Autobiography*, 12–23.

as alternatives for conscious choice, ends *are* future possible states of affairs in the respects that or insofar as they are understood . . . But one cannot understand the choice [among possible ends] . . . unless the ends are compared *with respect to choosing*, and a comparison of possible purposes in this respect is an evaluation. To understand some future possibility as a chosen telos is to affirm it as good and thus to compare it with alternatives in terms of better and worse.

At least implicitly, then, every activity understands itself in terms of some principle of purpose, that is, a principle of good purposes in view of which alternative ends are compared with respect to choosing.[31]

The upshot of this is that insofar as one affirms that humans exist with self-conscious awareness, as Darwin rightly does, one implicitly affirms in a transcendental sense that we always exist with some implicit background principle or conception of the good. This is one important example of what Gamwell means when he says that subjectivity necessarily includes certain inescapable understandings. To exist with self-conscious awareness is necessarily to exist with some implicit understanding of the good by which alternative possibilities are evaluated as better or worse. Thomas Nagel appears to offer a somewhat similar perspective when he declares that human action "is explained not only by physiology, or by desires, but by judgments. We are the subjects of judgment[s] . . . and those judgments have a subject matter beyond themselves. We exist in a world of values and respond to them through normative judgments that guide our actions."[32] It is these normative or evaluative judgments, Gamwell contends, that require and presuppose some implicit conception of the good—some evaluative yardstick by which alternative aims are judged better or worse as future possible states of affairs.

Following Whitehead and Hartshorne, Gamwell proceeds to argue that this implicit conception of the good, which is always part of the background of existence with understanding, is metaphysical in character. I will forestall a discussion of metaphysics until the next chapter, but here let me close by simply reiterating Gamwell's point that human subjectivity, what Darwin calls self-conscious existence, necessarily entails certain inescapable background understandings, which include some underlying conception of the good by which alternative possibilities are evaluated. What this

31. Gamwell, *Existence and the Good*, 101.
32. Nagel, *Mind and Cosmos*, 114.

suggests, therefore, is that Darwin's account of human existence implicitly affirms, in a transcendental sense, the inescapable presence of some conception of the good by which alternative possibilities are normatively assessed. Thus, once again, contrary to the prevalent, two-dimensional reading of Darwin, Darwin's thought implicitly presupposes a more inclusive and richer account of existence than is commonly assumed.

SUMMARY

In this chapter I have sought to articulate and unpack Darwin's key observation that what distinguishes human existence from that of other species is our capacity for self-conscious awareness. This capacity to remember the past, enjoy the present, and anticipate the future, is, as I've shown in Darwin's own life and scientific endeavors, a thoroughly teleological and evaluative enterprise. Moreover, as Gamwell makes clear, this capacity for subjectivity necessarily entails some dim background awareness of an evaluative standard by which alternatives are assessed. In short, I have sought to show that Darwin's affirmation of existence with understanding, which he rightly attributes to human beings, necessarily presupposes, in a transcendental sense, some form of teleology of value. As I will seek to argue in the next chapter, such a teleology of value, which makes sense of the grandeur rather than the vanity of the evolutionary process, can be given a sound metaphysical backing only by way of process or neoclassical metaphysics.

5

A CASE FOR GRANDEUR

Value and the Evolutionary Process

February 15, 2013, was something of a wake-up call for the world when a sizeable meteor stealthily penetrated the earth's atmosphere undetected by scientists only to appear suddenly racing brightly across the morning Russian sky at forty thousand miles per hour before unleashing a massive explosion twelve to fifteen miles above the city of Chelyabinsk. The blast, which contained energy equivalent to three hundred thousand tons of TNT, was "the largest explosion of its kind in more than a century." Thankfully, because the blast occurred so high above the city, the resulting structural damage and human injuries were not catastrophic. Had the meteor, which originated as an asteroid before entering the earth's atmosphere, hit the city directly, the outcome would have been cataclysmic. As scientists note, "giant impacts [in the past, larger than the Russian meteor,] have changed the course of life on Earth, notably 65 million years ago when an object several miles wide slammed off the coast of Mexico and killed off the dinosaurs."[1] Such apocalyptic-like events remind us of the contingency of human life and indeed of the evolutionary process itself.

In his 2011 film *Melancholia*, the Danish filmmaker Lars von Trier explores this relationship between calamitous natural events and the contingency of life, meaning, and value. Specifically, he ponders the possibility of an apocalyptic collision between earth and another massive space orb, which he calls the planet Melancholia. The film focuses on a contemporary

---

1. Chang, "Size of Blast."

family going about their normal lives with the looming threat of Melancholia moving toward the earth. One character, played by Kiefer Sutherland, is a secularistic man of science who tells his young son not to worry because the current projections estimate that Melancholia will travel closely but safely past the earth. Thus he encourages his son to view the growing image in the sky as an exciting adventure of astronomical observation. However, as the film progresses and Sutherland's character learns of revised estimates that predict an inevitable direct collision, he quietly retreats from his son, wife, and sister-in-law and goes off and commits suicide—leaving them alone, along with the totality of all life on earth, to face total annihilation. The film ends with his wife, son, and sister-in-law holding hands in a ritualistic circle of solidarity as Melancholia engulfs the earth in an act of cataclysmic nonbeing. Von Trier uses this scientific, albeit remote, possibility of massive cosmic collision and the total destruction of life on earth to ponder the powerful existential question of Ecclesiastes, namely, "Is it all hollow?" Is it all vanity? For von Trier, the answer is yes, for in the end, no value, no efforts, no acts of solidarity or scientific discoveries are retained: all is lost.[2]

Darwin himself reflected on this cosmic-existential question late in life. In his autobiography, he writes:

> With respect to immortality, nothing shows me how strong and almost instinctive a belief it is as the consideration of the view now held by most physicists, namely, that the sun with all the planets will in time grow too cold for life . . . Believing as I do that man in the distant future will be a far more perfect creature than he now is, *it is an intolerable thought that he and all other sentient beings are doomed to complete annihilation after such long-continued slow progress.* To those who fully admit the immortality of the human soul, the destruction of our world will not appear so dreadful.[3]

Contemplating this question through the prism of his then deepened agnostic outlook, Darwin laments the eventual loss of all that is gained through history and evolution, especially as it pertains to human achievement. That is to say, contemplating the question of the ultimate significance of evolutionary life in the "deep future," Darwin appears to side with von Trier in concluding that all is vanity because, in the end, all that evolution

---

2. Thorsen, "Longing for the End of All." Thorsen puts this question to von Trier. MIT physicist Max Tegmark contends that the question of whether the earth will be impacted by another massive and deadly asteroid is not a matter of *if* but only a matter of *when*: see Tegmark, *Our Mathematical Universe*, 374.

3. Darwin, *Autobiography*, 74–75 (italics added).

produces is lost.[4] This, I take it, points to the key underlying question: *Is what evolution produces, including human endeavors such as scientific inquiry, retained in the ultimate scheme of things, or is it all eventually lost in a cosmic etch-a-sketch—resulting in the same outcome as if it had never occurred at all?* Here, Darwin seems to imply that all is lost because there is no means or entity by which the values achieved in the course of events are retained. The crucial and precise issue is not, as he suggests, whether there is immortality of the soul, but rather whether the lives lived and the goods and values achieved in nature and history have everlasting or merely transient significance. In other words, when Darwin earlier proclaimed in *Origin* that there is grandeur in an evolutionary account of nature, that judgment presupposes, I will seek to show, an affirmation of everlasting rather than merely transient worth. My contention is that the Darwin who affirms the grandeur of evolution implicitly presupposes that the value and beauty achieved within nature and history have everlasting and not merely transient significance. This again is where Whitehead, Hartshorne, and Gamwell can be of help to Darwin. But to see this more clearly, let us begin by examining one notable effort to articulate a naturalistic conception of value.

A NATURALISTIC CONCEPTION OF VALUE

In his book *Meaning in Life: The Creation of Value*, MIT philosopher Irving Singer attempts to set forth a naturalistic understanding of value inspired by Darwin, Dewey, and others. Singer makes a fundamental distinction between "meaning *in* life" and "the meaning *of* life." He affirms the former but dismisses or at least downplays the latter. Western thought, he suggests, has been too preoccupied with this latter formulation, especially in the form of a question concerning the meaning of life. From the "traditionalist" perspective, the meaning of life has a fixed or prior content, "as if it were something preexisting" and "something *findable*" that we need to discover in God or the Platonic Good. Conversely, modern "absurdist philosophers," such as Camus and Sartre, claim that "the universe has no overarching

---

4. The apt phrase "deep future" comes from the Canadian philosopher J. L. Schellenberg. Schellenberg points out that evolutionary theory has taught us to take a very long view of time, what he calls "deep time," but such scientific thinking has generally focused only on the past, the "deep past." Yet, as Schellenberg rightly notes, deep time also applies to the very distant future—the "deep future." See Schellenberg, *Evolutionary Religion*, 3.

purpose." Hence, they "exhort us to live in a purposive manner that will be free of the delusive hopes about corroboration from the universe. They call this an acceptance of our basic absurdity," that is, the absurdity of our lives in a world without meaning.[5] Both of these views, Singer contends, focus on the wrong question. The universe as a whole may indeed be without meaning, but as creatures *within* the evolving world, we can and do pursue values and goods that sustain and enhance our lives—values that *we* find meaningful. Thus he declares:

> I therefore analyze meaning in life as a pattern of existence that engenders whatever values are required to go on living creatively. I suggest that this is what many, perhaps most, people who ask about a meaning *of* life are really concerned about. Even if there is no prior being, or category of being, that can yield an objective and independent meaning of life, living things have meaningful lives by creating values from within themselves. In that sense meaning in life *is* the creation of value.[6]

Singer appears here to affirm a teleology of value insofar as he affirms that humans and perhaps some other species pursue the creation of value in multitudinous ways. His central claim, however, is that such activities have value even if life or the universe *as a whole* has no meaning or value. But before addressing this larger issue, he makes a further distinction between happy or meaningful lives and "significant" lives.[7]

Given the fact that Singer tends to define meaning and value as a matter of preference, he suggests that a meaningful life or valuable activity is whatever a person deems it to be, that is, whatever one happens to desire or prefer: "Our purposes are directed toward the fulfillment of our desires and the acquisition of what we value." Therefore, he rejects the notion that genuine value or meaning must be defined in terms of ends that are comprehensive, lasting, or freely chosen. "The man whose waking hours are dominated by a compulsive need to work . . . or to seduce beautiful women, may insist that this gives meaning to his life. Are we prepared to say that it does not?" Singer apparently is not ready to make such a claim, for he proceeds to state, "What seems to one person like meaningless enslavement may well appear to another as the creative giving of himself." Yet, given his suggestion above that "meaning in life *is* the creation of value," neither

---

5. Singer, *Meaning in Life*, x, 40, 42, 37.
6. Ibid., xviii.
7. Ibid., 113.

is Singer fully content to embrace a relativistic outlook. "The pluralism I have been proposing may seem equivocal and overly relativistic to some people . . . Should we not distinguish between behavior that merely *seems* to be meaningful and behavior that really is?"[8] In one sense, he answers no: there is no valid way to evaluate the meanings or values individuals choose to pursue; there is no genuine evaluative yardstick by which to distinguish among better and worse aims or alternatives. However, in another sense, he is not quite comfortable with his own relativistic tendencies. Thus he proceeds to introduce a distinction between "significance and meaningfulness." As he puts it, "A significant life—one that is more than just happy or meaningful—requires dedication to ends that we choose *because* they exceed the goal of personal well-being. We attain and feel our significance in the world when we create, and act for, ideals that may originate in self-interest but ultimately benefit others." Echoing Darwin's notion of morality as a social instinct, Singer declares: "This mode of life comes naturally to us. It employs intelligence and imagination of a sort that is highly evolved in human beings." Sounding also like John Dewey, he further adds: "Throughout the varied pursuits that make a life significant, what remains constant is the growth of meaning in the service of transparent ideals . . . [W]hatever the activities we may prefer, we can recognize that the significance of any life will always be a function of its ability to affect other lives. And not that alone, since our perfectionism involves a longing to create the greatest possible good or beauty to which our imagination gives us access."[9] Finally he concludes by noting, "Whether we opt for a relativism [among values] or seek objective standards that reveal what is truly beneficial, a particular life can have significance only insofar as it augments the meaning and happiness of life *as a whole*, regardless of any effect upon one's own desires."[10] So one can live a happy or meaningful life without ever living a life of significance, but the latter, he implies, is more noteworthy and perhaps more honorable insofar as one pursues aims and ideals that creatively contribute to life as a whole.

Whatever may be the strengths or shortcomings of Singer's ethic and his distinction between meaning and significance, he is right to insist, first, that what ultimately measures significance is an action's contribution to life

---

8. Ibid., 106–112, 111, 112–13.

9. Ibid., 113, 115, 117; for an example of Dewey's thinking along these lines, see Dewey, *Common Faith*, 22–25.

10. Singer, *Creation of Value*, 117–18 (italics added).

as a *whole*; and, second, that the ultimate aim or ideal by which we measure what is truly good must have some dynamic aspect or quality, such that the ideal itself affirmatively includes that life is about the creation of value. These insights notwithstanding, my chief criticism is that Singer does not adequately answer the question of whether values and significant lives are ultimately lost in a cosmic etch-a-sketch, or whether they have lasting worth. For instance, in his discussion of happiness, he offers an extended reflection on mortality, life, and value:

> As persons who are happy, or hope for happiness, each of us wants to go on forever and we are saddened by the realization that this will not happen and may even be impossible. We may want to prolong our lives indefinitely, but we suspect that *everything in nature disintegrates and finally disappears*. Though we may cling to theories about life in another world after death, we also . . . fear that they are implausible. To some extent, everyone who thinks about the matter feels that his existence must be finite.
>
> If we had nothing else in consciousness, our fixation on our approaching doom would make life an unalloyed horror . . . Instead, *we mitigate the sentiment of dread* by fulfilling our nature not merely as persons who live or die within our separate being but also as expressions and embodiments of a life that includes more than just its particular manifestation in ourselves.
>
> In cultivating this further attitude, we move beyond our individuality and diminish our concern about its finitude. We expand our own selves by creating additional selves that issue from us: children who can live on when we are gone. We make material objects that will enter into the experience of others whether or not we are still alive. We create institutions and engage in pursuits, as in science or technology or the humanities, whose accomplishments endure long after we have died.[11]

Singer's argument, in outline, consists of five points: (i) We desire to live or "go on forever." (ii) Yet, we suspect that "everything in nature disintegrates and finally disappears." (iii) This realization of "approaching doom would make life an unalloyed horror." (iv) But we can at least try to "mitigate [this] sentiment of dread" by moving beyond ourselves and seeking to contribute to the lives of others through our relations, creations, and contributions. And (v) these contributions and "accomplishments endure long after we have died." Singer focuses here on the question of individual mortality and

---

11. Ibid., 119 (italics added).

seeks to "mitigate" or distract from this inevitability by suggesting that, even though each individual dies, s/he can contribute to the ongoing lives of others as an aggregated whole.

Let me offer two lines of response—one brief and one more extended. First, as we saw in the last chapter, humans exist with understanding; it is because we have the capacity for self-conscious awareness, as Darwin rightly noted, that we inevitably ask questions about the meaning and nature of existence. That is, we ask not merely about meaning *in* life but also about the meaning *of* life: we ask whether the *whole* of which we are a part itself has meaning and worth; we recognize our finite and fragmentary nature in relation to the whole of existence. Thus, one cannot simply dismiss the question of the meaning of life as superfluous or misguided (the way Singer does), for it is precisely because we have the capacity to conceive and evaluate the whole that enables us to conceive and value the parts. This was part of Gamwell's insight as outlined in the last chapter.

Second, as Darwin also implicitly recognized, the relevant question is not merely whether each individual perishes but whether the whole or totality, to which each individual contributes, itself perishes. As Singer himself says above, "everything in nature disintegrates and finally disappears." Thus, by his own account, this would apply not only to individual entities (both present and future) but to whatever conception of the whole he also presupposes. As he hints at a few pages later, when he says, "No ideal is ontologically supreme," he denies the enduring reality of any universal individual or ground of value.[12] In short, there is no genuine or enduring whole for Singer; what he means by the whole is merely an aggregate of perishing finite entities, none of which endures or retains any value. Thus, whatever we contribute to, whether it is other individuals, institutions, or the human race itself, eventually "disintegrates and finally disappears." Whether evolution produces various species or not or whether we build creative institutions or not, the oceans of time wash them away equally, just as if none of them had ever occurred. The physicist and cosmologist Brian Greene gives explicit voice to what is implicit here in Singer. Speaking of the notion of a "Cyclic Multiverse," Greene reflects: "In this scenario, the universe as we know it would merely be the latest in a temporal series, some of which may have contained intelligent life and the culture they created, but are now long ago extinguished. *In due course, all of our contributions and those of any other life-forms our universe supports would be similarly*

---

12. Ibid., 122.

erased."[13] Unlike Singer, Greene spells out here the implications of a cosmic etch-a-sketch, given the underlying assumption that there is no enduring or metaphysical whole. The "absurdist" philosophers, for their part, at least face this implication given their own underlying convictions. In contrast, Singer, who shares these same convictions, seeks to distract our attention away from this conclusion. Thus, his *meaning in life* is just a temporary form of medicating the underlying dread that Darwin and von Trier lay out in the open. Contrary to his intention, Singer, like von Trier, implies that it is ultimately indifferent whether we live lives of significance or self-interest because in the end all is lost. Singer comes closest to acknowledging this when he asks: "But if significance arises from pursuing an ideal that is itself variable and haphazard, or at least lacking in objective authority as far as the cosmos is concerned, why would anyone risk an iota of well-being in order to have a significant life? Why think that this aspect of our nature really matters? Indeed, why should we think that anything does?"[14] In spite of this moment of near clarity about the inescapable implications of his own view, Singer nevertheless keeps insisting that if it matters to an individual, if it contributes to some other individual or institution, then it has genuine worth. But in this insistence, he repeatedly commits the partialist fallacy insofar as he asserts that the part (the individual, institution, or species) can achieve lasting value and meaning even though the whole (the cosmos or whole of reality), which contains and includes the part, has no lasting value or significance.[15] Again, at least the "absurdist" philosophers acknowledge the incoherence of asserting meaning in the face of what Singer and they take to be ultimate meaninglessness. *For if what we do does not matter ultimately, then it does not ultimately matter what we do, whether we live lives of significance or not, because it all comes to the same end.* As Gamwell perceptively observes, it is "inconsistent to say that what we become could have significance without making an ultimate difference. Were there nothing ultimate at stake in what we do, then ultimately there would be nothing at stake. We who choose among our specific possibilities by taking some

---

13. Greene, *Hidden Reality*, 138 (italics added).

14. Singer, *Creation of Value*, 122.

15. The phrase "partialist fallacy" comes from Gamwell, by which he seeks to denote the fallacious attempt to affirm the moral worth or value of a part while concomitantly denying that the whole (inclusive of the part) has moral worth or value; see Gamwell, *Beyond Preference*, 12, 34.

purpose to be good cannot sensibly believe that this purpose finally has no point at all."[16]

We can likewise see the implication of this for Darwin's perspective. When he claims that there is grandeur in nature, he necessarily presupposes that what evolves or occurs makes some lasting and ultimate difference.[17] It is precisely Darwin and his evolutionary insights that rightly teach us to consider the deep recesses of time in terms of both the past and the future. Thus, we need a philosophical conception that enables us coherently to affirm the grandeur of evolution and all that it produces in the ultimate long run; anything short of that, we need to acknowledge that all is vanity, for all is eventually lost in the mists of time.

WILLIAM JAMES:
EVOLUTION AND THE QUESTION OF VALUE

In a prescient essay written just before the dawn of the twentieth century, William James insightfully grasped the set of issues at stake in this discussion of enduring worth. In seeking to analyze the pragmatic implications of various philosophic conceptions or worldviews, James notes that it is not so much the *retrospective* outlook that matters in testing alternative explanations of, for instance, how the world came to be, since both a materialistic and a theistic explanation lead to the same present moment; rather, it is their implications *prospectively* for how we interpret the meaning and significance of our present and future actions that really matters. In short,

---

16. Gamwell, *Politics as a Christian Vocation*, 104-5.

17. Likewise, when Darwin laments the loss of his beloved daughter Annie—"Oh, that she could now know how deeply, how tenderly, we do still and *shall ever love her* . . . [italics added]!"—he presupposes that this love, this worth, is enduring and not extinguished in the sands of time (Darwin, *Autobiography*, 102). On the surface, one might say that Darwin simply means that he (and his wife) will love Annie until he dies. But the implied meaning and sentiment, I would argue, are much deeper than that. He is not saying or implying, "I will love you until I die, but then that love is lost, extinguished, and dead, just as you are now." On the contrary, the power and depth of love and fidelity presuppose a deeper and everlasting affirmation of worth and significance, an everlasting bond of "creative fidelity," as the French philosopher Gabriel Marcel aptly describes it. Thus, Darwin is expressing here, I would contend, his deep trust in the *everlasting* worth of Annie's life and the love that they shared together. Marcel, throughout his many works, insightfully grasps the existential implications of our relationships, especially in terms of their implicit trust in and affirmation of everlasting worth; see, for instance, Marcel, *Creative Fidelity*.

James correctly recognizes that the persisting question about Darwin and his legacy is not the retrospective question of origins (Darwin was right on this front) but rather the prospective question of value. "Theism and materialism, so indifferent when taken retrospectively," says James, "point when we take them prospectively to wholly different practical consequences, to opposite outlooks of experience. For, according to the theory of mechanical evolution, the laws of redistribution of matter and motion, though they are certainly to thank for all the good hours which our organisms have ever yielded us and for all the ideals which our minds now frame, are yet fatally certain to undo their work again, and to redissolve everything that they have once evolved."[18] In seeking to further spell out the ultimate implications of this materialistic and mechanistic conception of evolution, James quotes the poetic Lord Arthur Balfour (1848–1930):

> "The energies of our system will decay, the glory of the sun, will be dimmed, and the earth, tideless and inert, will no longer tolerate the race which has for a moment disturbed its solitude . . . The uneasy consciousness which in this obscure corner has for a brief space broken the contented silence of the universe, will be at rest. Matter will know itself no longer. 'Imperishable monuments' and 'immortal deeds,' death itself, and love stronger than death, will be as if they had not been. Nor will anything that is, be better or worse for all that the labor, genius, devotion, and suffering of man have striven through countless ages to effect."[19]

This "is the sting of it," observes James; this is the pragmatic implication of materialism, namely, that "when these transient products are gone, nothing, absolutely *nothing*, remains, to represent those particular qualities, those elements of preciousness which they may have enshrined. Dead and gone are they, gone utterly from the very sphere and room of being. Without an echo; without a memory; without an influence on aught that may come after, to make it care for similar ideals. This utter final wreck and tragedy is of the essence of scientific materialism." Hence, "the true objection to materialism," adds James, "is not positive but negative." It is "the disconsolateness of its ulterior practical results" that matters; for it pragmatically

---

18. James, "Philosophical Conceptions and Practical Results," 70–71. James's essay was written in 1898 as a lecture he gave at the University of California, Berkeley.

19. Ibid., 71; James cites Balfour's 1895 book *Foundations of Belief*, 30.

undercuts any "permanent warrant for our more ideal interests" and values. It undercuts the worth and fulfillment "of our remotest hopes."[20]

To be sure, James recognizes that Singer and others may downplay the import of this conclusion by protesting that we should not take such a long view, that such concern with ultimacy is to ask about that which is "so infinitely remote as to mean nothing for a sane mind. The essence of a sane mind, [they] may say, is to take shorter views, and to feel no concern about such chimaeras as the latter end of the world." One notes, however, the irony of modern thinkers arguing against taking the truly long view since it was precisely Darwin and other evolutionary thinkers who correctly taught us to take the long view of time. For his part, James rightly replies that such attempts to delimit the human spirit, to put blinders on the scope and range of our capacity to envision the future, is to "do injustice to human nature ... The absolute things, the last things, the overlapping things, are the truly philosophic concern; all superior minds," he retorts, "feel seriously about them, and the mind with the shortest views is simply the mind of the more shallow man." On this question, he insists, "the positivists and [deniers] of metaphysics are in the wrong."[21]

James, like Darwin a few decades earlier, recognizes that the theistic options available in the late nineteenth century are less than optimal. Nonetheless, "The notion of God," James adduces,

> however inferior it may be in clearness to those mathematical notions so current in mechanical philosophy, has at least this practical superiority over them, that it guarantees an ideal order that shall be permanently preserved. A world with a God in it to say the last word, may indeed burn up or freeze, but we then think of Him as still mindful of the old ideals and sure to bring them elsewhere to fruition; so that, where He is, tragedy is only provisional and partial, and shipwreck and dissolution, not the absolutely final things.[22]

20. James, "Philosophical Conceptions and Practical Results," 71. In our 21st century context, *New York Times* science writer Dennis Overbye describes the view of Oxford philosopher Nick Bostrom along these same materialistic lines: "The best we [as humans] could have hoped for was to be another evolutionary phase in the zigzag development of earthly life on the way to who knows what. But in a few billion years, the sun will die, and so will the earth, and our descendents—if they are still on it. The universe will not remember us or Shakespeare or Homer" (Dennis Overbye, "A Case for Why We're Alone").

21. James, "Philosophical Conceptions and Practical Results," 72.

22. Ibid., 71.

Thus it is in these differing prospective outlooks on life, practice, and hope, James concludes, that the real meaning of and fundamental difference between theism and materialism is to be found. "Materialism means simply the denial that moral order is eternal, and the cutting off of ultimate hopes; theism means the affirmation of an eternal moral order and the letting loose of hope."[23]

Though James points toward a genuine insight here, he also contributes to the lack of clarity in theistic formulation. That is, he clouds matters by equating the notion of an ideal or eternal moral order with the notion of everlasting worth and significance. What is important here is not that God "guarantees an ideal order" in some prefixed, Platonic sense; this is the kind of static language and conception that Singer rightly criticizes. Rather, James's correct intuition is that what is achieved in life, nature, and history is "permanently preserved" in the divine life; thus, the world will "indeed burn up or freeze" at some point, but what has been achieved, however great or small, is not all for naught, as materialism inescapably implies; rather, there can be genuine grandeur in the process of evolution and extinction precisely because there is a universal, divine individual internally related to all that has occurred. What James is implicitly pointing toward here, and what Darwin was implicitly looking for decades earlier, is not a theism that offers a teleology of design, which Darwin and James both rightly reject, but rather a revised theism that can make sense of a teleology of value—one that can make sense of our pursuit of value in a fragmentary and transitory world. Here is where Whitehead, Hartshorne, and Gamwell offer Darwin, James, and us a way forward.

A PROCESS-METAPHYSICAL CONCEPTION OF VALUE

Within the limits of this study, I will not attempt a systematic account of Whitehead, Hartshorne, or Gamwell's metaphysics, which share much in common. Rather, by drawing upon their respective works, I will outline a way forward in terms of integrating our understanding of evolution with our understanding of value. This outline will consist of three parts: an organic rather than mechanistic view of nature; a process conception of God as an evolving universal individual or temporal series of all-inclusive wholes; and an integrated understanding of enduring worth, the divine life, and a teleology of value. In order concisely to introduce Whitehead's complex

23. Ibid., 72.

metaphysical thought, I will at times draw upon Donald Sherburne's masterful edited volume in which he thematically arranges and directly quotes from Whitehead's magnum opus *Process and Reality*.[24]

An Organic View of Nature

At the outset, I suggested that the nineteenth century failed to offer Darwin adequate philosophical or theological resources to help him integrate his evolutionary insights with his genuine sense of value, beauty, and worth. Whitehead himself attests to this fact when he remarks: "In the nineteenth century, some of the deeper thinkers among theologians and philosophers were muddled thinkers. Their [muddled thinking] was [caused] by incompatible doctrines," specifically, mechanistic determinism and freedom. Indeed, Whitehead contends, this tension between determinism and freedom lies at the base of modern Western thought and involves "two attitudes" that "are inconsistent. A scientific realism, based on mechanism, is conjoined with an unwavering belief in the world of men and of the higher animals as being composed of self-determining organisms. This radical inconsistency . . . accounts for much that is half-hearted and wavering in our civilization." Here Whitehead points to our inability to integrate our lived experience as self-determining individuals, who value, desire, and seek aims in the world, with our orthodox scientific doctrines that insist that there is only blind mechanism at work in nature—one in which the efficient cause wholly determines the effect—devoid of any form of final cause or teleology. Thus, "We are content," Whitehead continues, "with superficial orderings from diverse arbitrary starting points. For instance, the enterprises produced by the individualistic energy of the European peoples [such as Darwin creating photographs *for the sake of* studying the expression of human emotions] presuppose physical actions directed to [aims or] final causes. But the science which is employed in their development is based on a philosophy which asserts that physical [or efficient] causation is supreme, and which disjoins the physical cause from the final end. It is not popular to dwell on the absolute contradiction here involved."[25] As I have noted along the way, this disjuncture is evident in Darwin's own thought, given the incompatibility of his inherited mechanistic outlook with his implicit affirmation of a teleology of value. So, on the one hand, he denies any

24. Sherburne, *Key to Whitehead's "Process and Reality."*
25. Whitehead, *Science and the Modern World*, 82, 76.

notion of freedom, indeterminacy, or "chance" in nature when he dismisses such notions as simply "our ignorance of the particular cause" of each event while, on the other hand, he presupposes some degree of genuine freedom when he states that "man . . . can select, preserve, and accumulate the variations given to him by the hand of nature almost in any way which he chooses."[26] Even Paley's teleology of design, Whitehead observes, had fully bought into the dominant mechanistic mind-set; Paley simply insisted "that mechanism presupposes a God who is the author of nature." The only way of resolving this inconsistency and lack of integration at the heart of modernity, Whitehead concludes, is by recognizing that nature is not truly mechanism but rather organism.[27]

When one looks up the terms *organism* and *organic* in the dictionary, one finds definitions such as "Belonging to the constitution of an organized whole" (*organic*); "a whole consisting of dependent and interdependent parts" (*organism*); "Inherent in the organization . . . of a living being" (*organic*); and, even more philosophically, "The theory that in science everything is ultimately an organic part of an integrated whole" (*organism*).[28] These descriptions are in line with Whitehead's conception of the world in terms of a philosophy of organism. Instead of conceiving of reality as composed fundamentally of bits of inert matter or lifeless material, as Democritus did in antiquity and as scientific materialism has done in modernity, Whitehead envisions reality as consisting of micro-events, "actual occasions," or "drops of experience" that "prehend," "feel," or take account of the world in a nonconscious way. The emphasis here is on continuity and degrees of exemplification, which for Whitehead includes the continuity and differing degrees of exemplification among nonliving and living entities. "The philosophy of organism," declares Whitehead, "is a cell-theory of actuality. The cell is exhibited as appropriating, for the foundation of its own existence, the various elements of the universe out of which it arises. Each process of appropriation of a particular element is termed a prehension. I have adopted the term 'prehension,'" he notes, "to express the activity whereby an actual entity effects its own concretion [or taking account of] of other things." Specifically, "Every prehension consists of three factors: the 'subject' which is prehending, namely, the actual entity in which that prehension is a concrete element; the 'datum' which is prehended; [and]

---

26. Darwin, *Origin*, 173; Darwin, *Variation of Animals and Plants*, vol. 1, 3 [3/4].
27. Whitehead, *Science and the Modern World*, 76.
28. Brown, *New Shorter Oxford English Dictionary*, 2:219.

the 'subjective form' which is *how* that subject prehends that datum."[29] In a somewhat less technical manner, Whitehead summarizes his organic view of reality as follows:

> Each actual entity is conceived as an act of experience arising out of data. The objectifications of other actual occasions form the given data from which an actual occasion originates. Each actual entity is a throb [or drop] of experience including the actual world within its scope. It is a process of 'feeling' the many data, so as to absorb them into the unity of one individual 'satisfaction.' Here 'feeling' is the term used for the basic generic operation of passing from the objectivity of the data to the subjectivity of the actual entity in question. Feelings are variously specialized operations, effecting a transition into subjectivity. They replace the 'neutral stuff' of certain realistic philosophers. An actual entity is a process, and is not describable in terms of the morphology of a 'stuff.'[30]

By emphasizing that reality at root consists of organic process rather than material stuff, Whitehead points to at least three elements that are significant for our present study: first, that there are genuine internal as well as external relations within the world; second, that there is genuine novelty and creativity in the world; and, third, that wholes consist of societies of societies of interrelated entities.

Internal and External Relations

First, when Darwin talks about a tree of life or about evolution via descent, he is talking about how present entities are genuinely influenced by past actualities. That is to say, *what* presently exists *includes* in some way what came before it. This means that the present is *internally* related to the past while the past is *externally* related to the present: we inherit the past, the past does not inherit us; we are, in our very makeup, descendents of and thus affected by our familial and evolutionary ancestors; they are not affected by or descendent from us. Whitehead articulates this insight at the metaphysical level when he describes reality as a temporal series of emerging actual occasions that prehend the data of all past actualities. What the present occasion becomes is fundamentally informed but not wholly determined by the data of the past. Hence, the present inherits the past

---

29. Sherburne, *Key to Whitehead's "Process and Reality,"* 7, 8, 9.
30. Ibid., 8.

but does so as part of an evolutionary, organic process of becoming a new actual entity. Thus the evolutionary past is always included in the present as a factor of ongoing influence, to a greater or lesser degree. In brief, this organic view explains how and why the past informs the constitution of the present. In contrast, a materialist metaphysic, such as that found in Democritus or implied in some form by many modern scientists, presupposes that reality at its core consists of material bits that are only externally related to all other things: the bits themselves, however small, are never changed in their makeup or identity—they are eternal and unchanging; the only thing that changes is their external relations to other bits, in the form of various combinations or arrangements. Hence, the bits never evolve or inherit anything; they simply end up in different locations and in various combinations. One might be tempted to say that the combinations themselves evolve but the combinations themselves merely consist of temporary aggregates of the unchanging bits of matter; at root, there is no evolution here just unchanging bits in different locations. To borrow Dennett-like language, there are merely mindless bits in mindless combinations through and through. This is why Whitehead's philosophy of organism makes better sense of Darwin's evolutionary insights than does the materialist philosophy that Darwin himself inherited.

Novelty and Creativity

Second, by insisting that the data of past actualities inform but do not wholly determine present occasions, Whitehead's philosophy of organism explains why there is genuine novelty and creativity in an evolving world. As indicated above, each process of prehension consists not only of data from the past but also a subject that takes account of the data in its own subjective way. The influence of the past may or may not be massively dominant, but it is never total. This is why each actual entity is a new occasion, a new drop of experience that has never previously existed. On this account, Whitehead states:

> 'Creativity' is the universal of universals characterizing ultimate matter of fact. It is that ultimate principle by which the many, which are the universe disjunctively, become the one actual occasion, which is the universe conjunctively. It lies in the nature of things that the many enter into complex unity.

> Creativity is the principle of *novelty*. An actual occasion is a novel entity diverse from any entity in the 'many' which it unifies.[31]

Whitehead memorably sums up this point about the creative novelty of each new actual occasion: "The many become one, and are increased by one."[32] And this new one immediately becomes a datum for all future occasions, hence, a never-ending evolving process consisting of continuity, novelty, and change. This succinctly captures what evolution entails: without internal relations, there is no connection to the past; without genuine creativity and novelty, nothing new truly evolves or emerges—only the past reshuffled in different ways. If the evolution of species means anything, it means that novel forms of life come into existence, which were not wholly determined by the past. Whitehead's philosophy of organism makes better sense of this genuine novelty than does the dominant, mechanistic outlook in which every effect is wholly determined by its prior causes. To be sure, Whitehead acknowledges that mechanism can be a useful abstraction or shorthand, but it is a severely mistaken doctrine if one takes it as a fully accurate account of reality. Again, as he incisively puts it, "The aim of science is to seek the simplest explanations of complex facts. We are apt to fall into the error of thinking that the facts [themselves] are simple because simplicity is the goal of our quest. The guiding motto in the life of every natural philosopher should be, Seek simplicity and distrust it."[33] We must distrust the deterministic view of mechanism if we are to make full sense of the genuine novelty produced by organisms in an evolving world.

Organic Interrelations: Characteristics, Types, and Levels

And, third, as we saw in our definitions above, the notion of organism implies interrelation into a more inclusive whole or series of wholes. Our discussion requires some detailed unpacking and explanation for the sake of understanding this interrelation; thus, I will break up this section into subsections in order to help clarify matters.

An organism, as a living process, involves inheritance from the past, satisfaction in the present, and orientation or projection toward the future. In short, a living organism—as the history of a process—is a vector

---

31. Whitehead, *Process and Reality*, 21.
32. Ibid.
33. Whitehead, *Concept of Nature*, 91 (end of chapter 7).

or trajectory defined by a dominant characteristic. Each actual occasion emerges and perishes, becoming a datum for future actual occasions, but the life of a whole or persisting organism, which includes these actual occasions, is an ongoing trajectory. Whitehead calls this whole or trajectory, which is marked by a defining characteristic, a "society."[34]

Distinction between an Actual Occasion and a Society

According to Whitehead, it is critical that one understand the distinction between an *actual occasion* and a *society*. Actual occasions are the fundamental elements of reality (the micro-events at the root of all things), but the entities that endure or persist over time—the things that we are most familiar with in our everyday experience—are societies, not actual occasions. Societies include a series of actual occasions as their constitutive components, but the converse is not the case—actual occasions do not contain or include societies. As Whitehead reports,

> A society must involve antecedents and subsequents. In other words, a society must exhibit the peculiar quality of endurance [i.e. persistence over time]. The . . . things that endure are all societies. They are not actual occasions. It is the mistake that has thwarted European metaphysics from the time of the Greeks, namely, to confuse societies with the completely real things which are the actual occasions. A society has an essential character, whereby it is the society that it is, and it has also accidental qualities which vary as circumstances alter. Thus a society . . . enjoys a history expressing its changing reactions to changing circumstances. But an actual occasion has no such history. It never changes. It only becomes and perishes. Its perishing is its assumption of a new metaphysical function in the creative advance of the universe.[35]

Summarily put, an actual occasion, which is the most fundamental element of reality, is a nontemporal micro-event, one which becomes and perishes, whereas a society is a temporally and, perhaps, spatially ordered series or trajectory of actual occasions marked by a defining characteristic. Alternatively stated, a society involves the stretchedness of a series of occasions temporally (over time) and, perhaps, spatially (extended in space) whereas actual occasions themselves lack such stretchedness or duration.

34. Whitehead, *Adventures of Ideas*, 204.
35. Ibid.

To illustrate by way of analogy, think of a quality watch in which the second hand moves in distinct intervals rather than continuously: an actual occasion is that micro-event that occurs during the instant when the second hand does not move whereas a society is that duration of inherited movement from one moment to the next.

To clarify, then, a society necessarily involves *both* serially ordered relations (antecedents and subsequents make it an "enduring object") *and* "a common element of form," which "is the defining 'characteristic' of [that] society." In biology, for instance, a living plant or animal is a society that both exhibits some common characteristic and "enjoys a history expressing its changing reactions to changing circumstances."[36]

Two Types of Societies: Nonpersonal and Personal

Whitehead then proceeds to distinguish two types of societies: a nonpersonal society and a personally-ordered one. The former not only involves serially ordered relations of antecedents and subsequents but also includes contemporary actual occasions; that is, a nonpersonal society includes members that are contemporary with one another, such as a plant or animal body, i.e., the different parts of a plant or animal body exist contemporaneously with one another. A personal society, however, involves *only* serial relations of antecedents and subsequents. For instance, whereas the different parts of my body exist contemporaneously (my hands and feet both exist at the same time), I as a person or subject exist only from a prior moment to a subsequent moment, as a serially ordered set of relations, which we call an individual life. What Whitehead seems to have in mind with this distinction is that nonpersonal societies *may include* one or more personal societies as subsocieties that give organizing integrity and direction to the whole; that is to say, those societies that include a personal society are further governed by some defining center of activity. "When we survey the living world," observes Whitehead, "animal and vegetable, there are bodies of all types. Each living body is a society, which is not personal . . . [T]he lower forms of animal life, and all vegetation, seem to lack the dominance of any included personal society." Thus he remarks, for instance, that "a tree is a democracy," by which he means that it is a nonpersonal society fairly equally distributed in its function and not led by any organizing center. In contrast, "most of the animals, including all the vertebrates, seem to have

36. Whitehead, *Process and Reality*, 34; Whitehead, *Adventures of Ideas*, 204.

their social system dominated by a subordinate society which is 'personal.'" Hence, nonpersonal societies that include personally-ordered ones are those that are governed by some subsociety that centers and gives direction to the whole. Like Darwin, Whitehead is thinking here in terms of differing degrees of exemplification: a dog includes a personally-ordered society but a human individual, with a greater capacity for self-conscious awareness, can exemplify such personal ordering to a greater degree.[37] But, as Hartshorne rightly indicates, humans while sleeping are indeed closer to a nonpersonal society than to a personally-ordered one. "In deep, dreamless sleep," observes Hartshorne, "a human body is primarily a society of cells; it then, like a many celled plant, lacks a dominant [center or personal ordering]. In such a state a human soul, if this means a conscious individual, is in abeyance." In short, one should not underestimate the implications "of our nightly dip into an almost vegetative state." For this daily transition shows "the relativity of personal identity" insofar as we move back and forth each morning and night in differing degrees between a more personally-ordered society and one less so ordered. In both cases, nonetheless, the key point about Whitehead's notion of a society is "the way successive members inherit from previous members the society's 'defining characteristic.'"[38]

The Universe Consists of Societies of Societies of Societies

Like a set of Chinese nesting boxes or Russian nesting dolls, in which there is a series of more inclusive wholes containing smaller parts, the upshot of Whitehead's process philosophy of organism is that the universe consists of a nexus of societies of societies of societies at differing levels of integration and coordination. "The Universe," he proposes, "achieves its values by reason of its coördination into societies of societies, and in societies of societies of societies. Thus an army is a society of regiments, and regiments are societies of men, and men are societies of cells, and of blood, and of bones, together with the dominant society of personal human experience, and cells are societies of small physical entities such as protons, and so on, and so on."[39] Whitehead's insight is to recognize that reality consists of ascending and cascading levels of mutual immanence, integration, and process. The part influences the whole, and the whole exercises influence

---

37. Whitehead, *Adventures of Ideas*, 205, 206.
38. Hartshorne, *Insights and Oversights of Great Thinkers*, 41, 358.
39. Whitehead, *Adventure of Ideas*, 206.

on the part: hence an army influences the lives of individual soldiers, and, likewise, individual soldiers influence the performance and life of an army. My point, again, is that there is a teleology of value at work in this process, as empirically evidenced by humans and some other species, which means that there is a pursuit of value influencing the actions of societies of societies of societies and, thereby, influencing the whole. In order to adequately address the question of the whole, however, which ultimately raises the metaphysical question of God, it is important first to distinguish between differing degrees of integration and duration.

Relatedness: Fragmentary and Transitory vs. Holistic and Everlasting

When Whitehead speaks of societies relating internally to all past entities, he implies that there is a significant difference between relating in a *fragmentary* and *transitory* way and relating in a *holistic* and *everlasting* way.[40] Let me take up each of these contrasting elements in turn (*fragmentary* vs. *holistic* and *transitory* vs. *everlasting*). All ordinary societies—personal or nonpersonal—relate to the whole of their past and to their present environment in a vague, partial, and fragmentary way. For instance, think of Darwin's tree of life: we as humans inhabit one twig of one branch of one section of the whole tree; thus we directly inherit certain lines and are only indirectly or vaguely related to other areas of the tree, which are nonetheless still part of the same tree. Yet, as creatures with the capacity for self-conscious awareness, we can conceptually envision the whole tree as Darwin did, though we still do so in a fragmentary way (i.e., our knowledge of the whole is incomplete). But that there is a genuine whole we do not doubt, even though we do not and cannot know it holistically. In contrast, to relate internally to all things in a holistic rather than a fragmentary way is to include them in a supreme or all-inclusive manner. Thus to relate to the tree of life in a holistic way is to include all living entities, past and present, in their fullness rather than in vague or negligible outline.

In terms of the other contrast (*transitory* vs. *everlasting*), ordinary societies—personal or otherwise—relate to all other entities in a transitory or fleeting way. Singer captures this well when he states "that everything in

---

40. My distinctions here are informed by Schubert M. Ogden's discussion of "Logical-Ontological Type Distinctions in Outline: Ten Theses" in his currently unpublished Metaphysical Notebooks, February 2011.

nature disintegrates and finally disappears."[41] This is what it means for an individual or society to relate transitorily to all other entities: its integrative capacity comes to a temporal end. But this raises the very question of value and grandeur that is at the heart of this chapter and work as a whole. Whereas Singer implies that there is only transitory relatedness, and thus that all value is ultimately lost in a cosmic etch-a-sketch, Whitehead suggests that there is an everlasting relatedness in the most inclusive personally-ordered society of societies of societies of societies, which is what we mean by the concept of God. "By reason of the relativity of all things," says Whitehead, "there is a reaction of the world on God. The completion of God's nature into a fullness of physical feeling [or prehension] is derived from the objectification of the world in God. He shares with every new creation its actual world; and the concrescent [or actualized] creature is objectified in God as a novel element in God's [prehension or] objectification of that actual world."[42] To get at this point more fully, it helps to shift from Whitehead's discussion of societies to Hartshorne's language of individuals.

The Universe as Composite or Compound

Whereas Whitehead distinguishes nonpersonal and personally-ordered societies, Hartshorne distinguishes "composite individuals" and "compound individuals." That is, compound individuals (like nonpersonal societies that include a personally-ordered subsociety) are those "colonies" that "involve a dominating 'personal' unit" whereas composite individuals lack such an organizing center.[43] The key question implied above is whether the universe is merely a composite—that is, an aggregate of fragmentary and transitory individuals that lacks both holistic and everlasting integration, *or* whether the universe is a compound individual that holistically and everlastingly integrates the world and the values it produces. Hartshorne formulates and responds to this query as follows:

> The first great problem of metaphysics was . . . that the universe is a single existent, while it also has as its parts all other existents. The answer to the question, how can this be? is the answer to the question, what do we mean by God? *For God is the compound individual who at all times has embraced or will embrace the fullness of*

---

41. Singer, *Creation of Value*, 119.
42. Whitehead, *Process and Reality*, 345.
43. Hartshorne, *Whitehead's Philosophy*, 57.

all other individuals as existing at those times. He is the only eternal (primordial and everlasting) individual, and the only one whose prehensions of others involve impartially complete vividness for all, wherever they may be in space or (past) time.[44]

Hartshorne proceeds to ask, how do we know that God exists? While he addresses this question in detail elsewhere, here he offers a few brief arguments.[45]

The Universe as Compound: Arguments for the Divine Existence

Hartshorne proposes that "the universe must have some . . . everlasting character, as the ultimate subject of change." Since the past is always inherent in the present, there must be some entity or subject of change that inherits *all* of the past and not merely some part of it. For instance, as fragmentary individuals, we as human beings inherit the past in a vague and partial way, and thus we know and remember just a fragment of it, such as our partial knowledge of the tree of life, but we rightly never assume that this means that there is no more to the past or to the tree of life than the fragment that we know or will ever know. Hence, there must be some subject or individual that inherits the past holistically and everlastingly rather than merely fragmentarily and transitorily. Otherwise, only part of the past would exist instead of all of the past. In short, the past, whatever it entails, includes *all* of the past and not merely *some* of it. In Hartshorne's words, "The past being immortal, there must be a complete cosmic memory, since the past in the present is memory."[46] This complete cosmic memory is found in the one divine individual.

Hartshorne offers another line of reasoning that directs our attention back to the question about the everlasting nature of value. All "action," he contends, "implies the faith that at no time in the future will it ever be true that it *will* have made no difference whether the action was well motivated or ill (successful or not)."[47] Like James and Whitehead, Hartshorne rightly

---

44. Ibid., 60 (italics added).

45. Ibid. For Hartshorne's extended discussion of theistic arguments, see, for instance, Hartshorne, *Logic of Perfection*, and Hartshorne, *Natural Theology for Our Time*.

46. Hartshorne, *Whitehead's Philosophy*, 60.

47. Ibid. This was the point I made earlier about Darwin's affirmation of undying love for his daughter Annie: in the face of her death and in the midst of his grief, Darwin trusts that this love will not be lost or nullified in the ultimate passing of time.

Darwin in a New Key

maintains that the question of value ultimately presupposes that what is actualized in evolutionary history makes an everlasting and not merely a transitory difference.[48] "I have yet to be convinced," says Hartshorne,

> that human beings can really demonstrate their belief in an absolute renunciation of everlasting (however modest) significance for their having lived as they in fact have. They can, I am confident, believe that their individual experiences will all be contained in their earthly lives from birth to death; but not that the eventual significance of these lives may one day reach zero. This is sayable in words, but what is the mode of living that expresses believing it? I hold with James and Peirce that the test of a belief's genuineness is the possibility of acting upon it. In this case I see no such possibility. One may say that the practical conclusion is, nothing matters. But this is mere words; while we live we show that for us something does matter.[49]

It is important to recognize here that Hartshorne is arguing for the everlasting significance of our finite and mortal lives, not for our eternal or ongoing existence. Neither we nor any other creature goes on having new experiences beyond death; all of our individual experiences occur within our "earthly lives from birth to death." As Hartshorne proceeds to say: "I agree with . . . Heidegger [and others] . . . that it is death that gives a human life as a whole its definiteness." So the issue at stake is certainly not whether we are mortal, for we are indeed mortal, finite, and definite. The

---

48. This is another way of stating Hartshorne's point about all of the past being included in the present; that is, to insist that the difference a human action makes is an *everlasting* difference is to express a special case of the metaphysical condition that *all* of the past is *always* in the present.

49. Hartshorne, *Insights and Oversights*, 360. When one thinks of a life that believes its choices and actions will make no ultimate difference, one thinks of Camus's character Mersault in *The Stranger* (alternatively translated as *The Outsider*). Mersault is indifferent to life's choices—e.g., whether to marry or not—because he believes that all choices are arbitrary, and he lives out such a life with brutal honesty. He shares Socrates's honesty, but he has a very different metaphysical, existential, and ethical outlook: whereas Socrates (Plato) believes that life exists in relation to the Good, and thus that one should devote one's life to pursuing questions of truth, justice, and integrity, Mersault believes that life is an absurd given in the midst of cosmic nakedness; we have freedom but without any metaphysical structure of meaning and value. Hence, one choice (to shoot or not shoot on the beach) is as arbitrary as any other choice: i.e., no choice will make any ultimate difference. Hartshorne's point is that no one actually lives like Mersault; no one actually lives and acts as if their choices in life are arbitrary or indifferent in the long run, in spite of what they may verbally say.

real issue, Hartshorne continues, is "what does our being definite matter if the universe has no way to retain the definiteness?"[50] In other words, if the universe has no way of retaining the particularity that evolves, if all that is achieved is eventually lost in the deep recesses of time, then in what coherent sense can one say that any action, life, or achievement has value or significance? To be sure, Hartshorne notes, one can verbally say that there is meaning and value in life while concomitantly denying that the universe retains any of this value in the long run; but saying it does not make it the case. In fact, I would add, there are conceptual as well as pragmatic reasons for arguing that it is not the case. Conceptually, the notion of value, by definition, necessarily implies some state of affairs that is a nonnullity: value is something not nothing. As indicated above, the question is not whether one can add zero to some value: concrete values, such as a human life, are definite; thus there comes a point when nothing further is added to them, i.e., we die. Rather, the real issue is whether a value ultimately becomes zero or sheer nullity: does it make sense to say that there are values in life but that in the long run, those values become zero and thus become null and void? I would argue no. For instance, in arithmetic terms $6 + 0 = 6$ (its definiteness is retained but nothing further is added to it; a life lived is ended but the value of its definiteness is retained) but $6 \times 0 = 0$ (*there is no value in nullity; any supposed affirmation that ultimately comes to zero is no affirmation at all*). This is why it does not make conceptual sense to say that value is real in the short term but nothing in the long term; when anything is multiplied by zero, there is only zero. Alternatively stated, *value presupposes difference, and nullity knows no difference*. This is why, on pragmatic grounds, one cannot truly live, act, or believe "that the eventual significance of these lives may one day reach zero."[51] To affirm value, such as Singer's notion of significant lives or the worth of scientific understanding, is to affirm that some desired state of affairs makes a genuine difference, but there is no genuine difference if all becomes null and void. This again is why Gamwell is correct to point out that it is "inconsistent to say that what we become could have significance without making an ultimate difference. Were there nothing ultimate at stake in what we do, then ultimately there would be nothing at stake . . . So far as I can see," adds Gamwell, "it is finally impossible to understand ourselves in terms of a final nullity, as if our lives were only a brief flicker of light with only darkness before and darkness after . . .

---

50. Hartshorne, *Insights and Oversights*, 360.
51. Ibid.

But if this is so," he concludes, then "there must be some everlasting reality to which [all] life makes a difference, and that conclusion points the way toward an all-embracing . . . divine individual."[52] What Gamwell indicates here is not only that value must make some ultimate difference, and thus not become null and void, but that there must be some everlasting entity for which it makes a difference: for the only way to make a difference is to make a difference to something that can receive or be affected by the change. And the only entity that could receive or be affected by *all* of evolutionary life and history, and thus for which all of life and history could make an ultimate difference, is a divine individual. What is required, therefore, is a truly adequate and coherent conception of divinity.

A Process Conception of God

The conception of God that Darwin rightly became increasingly skeptical of, and the one that Paley and others advocated, is one in which God relates merely externally and mechanistically to the world. This traditional view conceives of God as the author and designer of the world who mechanistically causes the world to come into existence—either directly in a creationist way or gradually through some teleology of design. Such a conception holds that God is not internally or socially related to the world and, thus, is not affected or changed by it. As Hartshorne remarks, "mechanism, materialism, and absolutism [applied either to God or the world] can all be viewed as special cases of the same error, [which is] the arbitrary reduction of one or more aspects of sociality to zero." In brief, it is largely because of the cultural dominance of this particular theistic conception that the debate concerning Darwin's work has focused for more than a century-and-a-half on the backward-looking question of origins: either God was the mechanistic cause of the world or the world was caused by blind mechanistic algorithms. But as James indicated more than a century ago, and as I have argued here, the real question pertains to the forward-looking question of enduring value and to the ultimate significance of whatever values are actualized by human beings and other species in evolutionary history. The process conception of God that Whitehead, Hartshorne, and others developed in the twentieth century, one that emphasizes God's internal and all-inclusive relations to the world as much as any external relations, is one that Darwin regrettably never had a chance to consider. At the heart of this

52. Gamwell, *Politics as a Christian Vocation*, 104, 105.

process view, which Hartshorne refers to as "neoclassical theism," are the notions of divine *relativity* and *dipolarity*.[53] Let me delineate each of these in turn.

Relativity

In the sense that matters here, to be relative to something is to be affected by it and thus, in that sense, dependent on it. Otherwise put, relativity implies internal relatedness: to be relative to something is to be internally related to it and thus affected by it. We are relative to and dependent on the past insofar as we are internally related to it; we are descendents of our ancestors, they are not descendents of us. Inheritability, i.e., the capacity to inherit, is an indicator of internal relatedness. Now, as discussed above, we and all other creatures relate to the past in a partial way, and thus we inherit the past in a fragmentary and vague manner. In contrast, divinity refers to the one compound universal individual that relates to the past and all entities in a supreme or holistic way. This is what Hartshorne means when he speaks of the divine relativity or sociality. God "is socially aware of all beings," says Hartshorne, "the actual as actual, [and] the possible as possible." He refers here to the distinction between present or past actuality versus future possibility: God does not confuse past, present, and future. That is, God, as internally and holistically related to all of the past, knows the past and present (which immediately becomes the past) as actual whereas the future, as indeterminate possibility, is a matter of probability not actuality. Hence, what has evolved God knows with full clarity, but what will evolve is a matter of uncertainty, probability, and contingency. For Hartshorne, all individuals, including the divine individual, are temporal, i.e. their existence is actualized in time; they endure or persist over time. Thus what distinguishes God from creatures is not temporality but the difference between supreme or everlasting temporality and partial or transitory temporality. We and all other creatures endure for a season and then die; God endures everlastingly, which is to say that God's existence is necessarily actualized in some contingent manner in relation to some evolving world or state of affairs. God is the all-inclusive individual or series of wholes that increases by way of addition. As Hartshorne observes, change occurs "in the form of addition of new terms with their new relationships. True, the past must then still exist in the depths of the present; but this does

---

53. Hartshorne, *Divine Relativity*, 28, viii.

not contradict the past's distinction from the present. For a constituent of a whole is not identical with that whole . . . [To say] that the event is 'no longer present but past' means that now a new and more inclusive whole possesses it as a part." Like a rolling snowball, the present includes the past as a new and ever-changing and ever-growing whole. "Our human consciousness has, of course, but a feeble direct awareness of the inclusion of the past in the present, via memory. But human direct awareness is not the measure of reality," Hartshorne concludes; the all-inclusive and everlasting divine individual is.[54]

Dipolarity

The notion of divine relativity emphasizes that God is internally related to the world in a supreme, holistic, and unending temporal manner. The conception of God and all actual entities as dipolar provides the conceptual framework for making sense of both external and internal relations. External relations, again, are those relations in which something is unaffected or unchanged by that relation. So, as said, the past is externally related to the present and thus unchanged and unaffected by the present. To be sure, present understanding of the past may change, but whatever occurred in the past is unchanged, which is what we mean by calling it past. Likewise, as we have discussed, all enduring entities are temporally-ordered series or trajectories marked by some defining characteristic or characteristics. These defining attributes or characteristics remain unchanged, either relatively or supremely, whereas the concrete experience of a society or individual is ever-changing as new occasions are added to its life. What I am getting at here and what the notion of dipolarity suggests is that all enduring entities have both an *abstract* and a *concrete* aspect and, thus, are dipolar. The abstract pole contains those aspects or characteristics of an entity that remain consistent over time, such as identity, whereas the concrete pole involves those new drops of experience that constitute the actual living of an organism. So, for example, a species is an abstraction that remains relatively unchanged over vast stretches of time; such an abstraction is one of the defining characteristics that identify a particular member of that species. That particular member is also identified by other characteristics, such as in the case of humans by what we call personality. Hence, my identity as a member of the human species remains unchanged throughout

54. Ibid., 37, 69.

my lifetime, and my personality remains relatively unchanged as well, but my experience changes from moment to moment as new actual occasions are added to the whole of my existence, which is what we mean by a *living being*. Thus, the whole of who I am, as both abstract consistent characteristics and as changing concrete, lived experience, is an ongoing series of new wholes throughout my lifetime; each new whole includes the past as well as the addition of new concrete, lived experience, e.g., $I_1 \to I_2 \to I_3 \to I_4 \ldots$ (I_4 includes the whole of my past plus the new lived experience that distinguishes it from I_3 and all previous wholes or actualizations of I). So, again, what distinguishes God from ordinary creatures is not that God is singular or monopolar (this is the critical mistake of traditional or classical theism), but rather that God is supremely dipolar (a trait Hartshorne calls God's "dual transcendence"). Ordinary creatures are dipolar merely in a fragmentary and transitory way. Thus, for instance, my abstract characteristics are merely relatively unchanging (personality can be lost, as the heartbreak of Alzheimer's makes all too painfully clear, and even the notion of species is only relative, as Darwin correctly pointed out) whereas in the case of God they are absolutely or supremely unchanging, which is what it means to say that God is the one individual that exists metaphysically. On this point, Hartshorne writes:

> The idea of God is a metaphysical idea. By this I mean that its definition employs only categories, concepts of strictly universal significance, claiming application to any and every conceivable existence. The defining characteristics distinguishing God from any other being, actual or conceivable, can be specified by such concepts. This is not true of any other individual being.[55]

Just as God is supremely absolute or unchanging in terms of the abstract characteristics that define any individual, so too is God supremely temporal and relative in terms of all-inclusive internal relatedness to all concrete actuality. So God's identity as God is externally related to the world and thus is unchanging, but God as internally related to the world is supremely receptive to all temporal change, and thus God as a whole, as the divine individual, is an everlasting series of changing wholes as the evolving world continually adds new concrete actuality to the unending divine life; the divine life evolves in relation to an evolving world. The great advance of a dipolar understanding of God and world, Whitehead observes, is that it coherently captures and expresses our correct "intuition of permanence

55. Hartshorne, *Creative Experiencing*, 93.

in fluency and of fluency in permanence ... In this way," he reflects, "God is completed by the individual, fluent satisfactions of finite fact, and the temporal occasions are completed by their everlasting union" in the all-inclusive and integrated divine life.[56]

Integration of Enduring Worth, the Divine Life, and a Teleology of Value

Having now outlined an organic view of nature and a dipolar conception of divinity, we are in a position to get to the crux of the matter in terms of how a process understanding can make sense of evolution and the question of value. Up to this point, I have argued that Darwin implicitly affirms a teleology of value insofar as he explicitly affirms domestic selection, the pursuit of beauty by humans and some species of birds, and a quest for understanding, such as exemplified by science. Moreover, he famously claims that there is grandeur in nature when understood in evolutionary terms. Grandeur rather than vanity, I have held, presupposes that whatever value and beauty are generated through evolutionary change and by the activities of creatures must not be lost in the deep recesses of time but rather must make some ultimate difference to an everlasting, divine individual. Let me now spell out in greater detail what I mean by a teleology of value and how it contributes to the genuine grandeur of the evolving universe.

From the outset, I have maintained that there is a critical distinction between a teleology of design and a teleology of value. A teleology of design holds that the world we know was in some way prefabricated to end up as it is. Summarily stated, a teleology of design presupposes some version of determinism, usually mechanistic, that controls events leading to a preset outcome. As I have indicated, the theistic mechanism of Paley's teleology of design is just as problematical as the mindless algorithmic mechanism of Dennett's version of Darwin. All forms of mechanistic thinking, theistic

---

56. Whitehead, *Process and Reality*, 347. It should be noted that one of the differences between Whitehead and Hartshorne is that Whitehead conceives of God as an unending or nonperishing *actual occasion*, and thus in this sense God is nontemporal, whereas Hartshorne conceives of God as an everlasting temporal compound *individual*—as a temporal series of wholes inclusive of differing actual occasions. Hartshorne thinks that his formulation makes better sense of Whitehead's overall view and, as suggested by my exposition to this point, I concur with Hartshorne in this matter. For Hartshorne's discussion, see Hartshorne, *Whitehead's Philosophy*, 83–90 where Hartshorne discusses "Whitehead's Idea of God" in relation to the question of temporality.

or otherwise, are an acid that dissolves our understanding of the genuine contingency, creativity, and value of an evolving organic world. But to reject mechanism or a teleology of design does not require us to deny the vivid evidence from our lived experience of the teleological pursuit of value in our everyday activities. Domestic selection, as Darwin took for granted, is a teleological act of breeding plants or animals for the sake of some desired aim. For instance, we now have the goldendoodle (a cross between a golden retriever and a poodle) because people desire pets with certain valued characteristics. I cannot think of a more self-evident example of a teleology of value at work in the world than this: goldendoodles would never exist were it not for the intentional pursuit of desired ends. But part of Darwin's insight was to recognize that a pursuit of value (such as beauty) is not limited only to the human species. Thus, as Whitehead again observes, there is a threefold urge in nature: to live, to live well, and to live better. Species, as he notes, not only passively adapt to their environments, but also actively modify them to their own ends. "In so acting, they are transforming the environment for their own purposes."[57]

But if there is indeed a teleology of value at work in the world, which is most explicitly evident in the human species, and if the meaning of value ultimately presupposes an everlasting, divine individual, then: What do we mean by value? What is the role of God in relation to the creation of value? And what is the role of God in relation to the enduring worth of value? We have touched on this third question, but we will return to it once we have addressed the other two.

Value

Given an organic rather than mechanistic view of nature, all actual occasions generate value insofar as they prehend, synthesize, and integrate the data from the past in their own act of self-creation; in self-creating, each actual occasion generates some (slight or significant) degree of novelty. This is why Whitehead describes creativity as the preeminent metaphysical principle, as the universal of universals. "To experience [or prehend] at all," Hartshorne declares, "is to create value." Hartshorne unpacks this as follows:

---

57. Whitehead, *Function of Reason*, 8, 7.

> On very low levels of actuality, as in atoms or light rays, there is extremely slight novelty (hence the relatively sharp predictions of physics, especially when dealing with large collections of similar actualities) and there is presumably equally slight value. In plant cells and the lower animals there is more novelty and value. In our species, and . . . in other primates and in whales, there is immensely greater creative power and greater value. The felt harmonies are more complex and rich. The value of harmonious feelings is proportional to their intensity, and this seems to depend upon the depth and variety of the contrasts in the data . . . The higher animals, especially humans, enjoy a rich life of thought that adds new dimensions to the contrasting factors of experience.[58]

Let me offer here three important points of clarification. First, what defines value is the richness of the integrated contrast between diverse and unified elements. Like Aristotle, who suggested that beauty involves the maximal integration of breadth and unity, Whitehead and Hartshorne define value aesthetically in terms of the process of maximizing unity-in-diversity, synthesizing diverse contrasts.[59] This is what Hartshorne means above when he says: "The value of harmonious feelings is proportional to their intensity, and this seems to depend upon the depth and variety of the contrasts in the data." So, greater richness in depth, unity, and contrast defines greater value while, conversely, the total lack of unity (chaos or anarchy) or the total lack of diversity (homogeneity) defines disvalue. On this point, Whitehead once memorably defined evil as homogeneity, by which he means a state of affairs that lacks or seeks to remove all diversity and contrast.[60] One need only think of twentieth-century fascism and its attempt to homogenize European societies to confirm Whitehead's definition. Second, following from this, not all value is positive or creative in the sense of adding enrichment to the organism or good to the world. This is not Leibniz or Dr. Pangloss speaking of the best of all possible worlds. Yes, to exist at all involves the act of prehending and synthesizing the world and, thus, the creation of some degree of value. Hence, "just to be a complex viable organism is an immense harmony in itself," observes Hartshorne. But this does not mean that all that is prehended is maximally valuable or good. "A suffering animal achieves some value; otherwise the will to live

---

58. Hartshorne, *Creative Experiencing*, 130, 129.

59. See Aristotle, *The Politics*, book VII, sect. iv, 1326a35 where Aristotle discusses beauty in relation to defining the most beautiful or good state.

60. Lowe, *Alfred North Whitehead*, vol. 1, 136.

would lapse. But the animal does not, while suffering, achieve the optimal value possible for its kind of creature. Also a creature may achieve the value it does enjoy in such a way as to limit disproportionately and tragically the value opportunities of other creatures," such as illustrated by invasive species.[61] And, third, when Hartshorne and Whitehead speak of lower or higher organisms, they certainly do not mean a teleology of design, as if the lower organisms are created for the sake of the higher. Instead, what they refer to are the varying capacities of different entities to integrate and synthesize diverse contrasts of data. This applies both to differing capacities to appreciate beauty and to differing capacities to suffer. In regards to beauty, for instance, Darwin takes this distinction for granted when, as we saw in chapter 3, he affirms the capacities of various birds to appreciate the beauty of certain forms and contrasts of color and sound, but he also declares that "obviously no [bird or] animal would be capable of admiring such scenes as the heavens at night, a beautiful landscape, or refined music; but such high tastes . . . depend on [a capacity for] complex associations."[62] But just as the facility to appreciate beauty varies, so too does the capacity to suffer: a rock's capacity to suffer is virtually nil, a plant or flower's is greater but still less than a sentient creature's, and more complex sentient creatures, such as humans, have a still greater potentiality to suffer, as illustrated by Darwin's anguished words over the loss of his beloved Annie.

God and the Creation of Value

In a teleology of design, the outcome is guaranteed in advance because God mechanistically determines or prefabricates the result. In contrast, in a teleology of value, divinity acts as a cosmic lure or "appetitive vision" that seeks to attract all actual occasions and enduring individuals toward greater unity-in-diversity. But precisely because all entities have some degree of creativity or freedom, the role of divinity is one of inviting influence, not dictatorial control or determinism. As noted earlier, one of Whitehead's important contributions is to recognize both ascending and cascading forms of influence—how each part influences the whole and how the whole influences each part. God, as the cosmic or universal individual supremely related to the world, provides the ultimate background influence for all entities. As Whitehead puts it, "God's nature is a primordial datum

---

61. Hartshorne, *Creative Experiencing*, 130.
62. Darwin, *Descent of Man*, 116.

for the World." This divine nature includes the potentiality of an evolving world in the form of the indefiniteness of possibility related to each new actual occasion. "Viewed as primordial," Whitehead remarks, "[God] is the unlimited conceptual realization of the absolute wealth of potentiality. In this aspect, he is not *before* all creation, but *with* all creation."[63] This distinction succinctly captures the crucial difference between a teleology of design and a teleology of value: the former envisions God as *before* all creation as the designer and prefabricator whereas the latter conceives of God as *with* all entities in their ongoing and evolving acts of greater or lesser creativity. Yes, God always invites the world toward greater creativity through the divine subjective aim of maximizing beauty or unity-in-contrast, which is an ever-present background influence for all actual entities, but whether and how the process of cosmic, natural, or human history in fact evolves toward greater beauty, value, and unity-in-contrast is a matter open to some degree of contingency, freedom, and uncertainty. If God is the *unmoved mover* for Aristotle that, like a magnet, exercises teleological attraction on the world but is totally unaffected by it, then God is the *moved mover* for Whitehead and process thought that, like a poet, is both influenced by and exercises influence on the world. "God's role is not the combat of productive force with productive force," reflects Whitehead; rather "it lies in the patient operation of the [influential] rationality of his conceptual harmonization. He does not create the world, he saves it: or, more accurately, he is the poet of the world, with tender patience leading it by his vision of truth, beauty, and goodness."[64]

God and Enduring Value

As revealed here by Whitehead's incisive statement that God "does not create the world, he saves it" *by receiving all of it*, the seminal contribution of neoclassical theism is that it focuses philosophy on the forward question of the ultimate worth of a dynamic and evolving world and not just on the backward issue of origins. Throughout most of Western history—whether it was Plato's idea of the Good, Aristotle's unmoved mover, the Stoic notion of a holistic cosmos diffused with a rational divine logos, a Christian conception of God as creator, Spinoza's one substance expressed through infinite attributes and modes, or a modern secularistic understanding of

---

63. Whitehead, *Process and Reality*, 348, 343.
64. Ibid., 346.

an evolving world determined by mindless mechanism—the question of what value or difference life ultimately makes has largely been ignored or denied. And this is because for most of Western thought there has been nothing to receive the world: either there has been a denial of any ultimate or metaphysical reality at all (modern secularism), or ultimate reality has been understood to be merely externally related to the world and thus not genuinely affected by or receptive to it (e.g., the Platonic Good, Aristotle's Unmoved Mover, the Stoic Logos, the traditional Christian God, or Spinoza's One Substance). As Gamwell nicely puts it, "to understand God as eternally complete is finally to deny any ultimate importance to what we do or become. Whatever effect we have in the world, we cannot make a difference to something that cannot receive it." The decisive contribution of process or neoclassical theism is that it articulates how the divine individual is internally related to the world, and, thus, how God is "the ground of all worth." As Gamwell further notes, "The effects we have in the world would be without any worth except that the world has the significance God gives to it by receiving all of it into the everlasting divine life. In the end, the only difference we can make is the difference we make to the divine good."[65] As we have seen, this is by no means to deny the contingency, creativity, novelty, and freedom of the evolving universe. But it does explain how one can coherently affirm the grandeur rather than vanity of such a universe in which all concrete actualities, from the smallest creatures to the largest suns, eventually perish. The divine receptivity of the world is what Whitehead calls its "objective immortality" in the "consequent nature of God." It is this objective immortality, as prehended and integrated into the holistic and everlasting divine life, that saves the "'perpetual perishing'" of the evolving world and thus defines it as a world of enduring worth and value.[66] In short, it is the difference between the genuine grandeur of an evolving world and the vanity of a cosmic etch-a-sketch.

CONCLUSION: DARWIN, VALUE, AND THE GRANDEUR OF AN EVOLVING WORLD

I began this study by suggesting that our twenty-first-century culture has yet fully to integrate Darwin's insights with our understanding of the world because we have failed to integrate the question of evolution with

---

65. Gamwell, *Politics as a Christian Vocation*, 104.
66. Whitehead, *Process and Reality*, 347.

the question of value. The chief reasons for this failure have been multiple. They include a misguided religious resistance to Darwin's understanding of natural selection and a truncated, two-dimensional interpretation of Darwin by most neo-Darwinists. Over against these prevailing voices, I have argued that the way forward is found in a three-dimensional reading of Darwin, which is premised on a critical distinction between a teleology of value and a teleology of design. A three-dimensional account recognizes not only the critical roles of survival and reproduction (as well as natural and sexual selection) but also the genuine role that the pursuit of value plays in the world. As evidenced most explicitly in the human species, when conditions permit, we survive and reproduce in order to pursue and enjoy the worth and enrichment of love, beauty, justice, and other forms of satisfaction: we survive and reproduce in order to live; we don't live merely in order to survive and reproduce. In essence, my hermeneutical conclusion from studying Darwin is that the question of evolution can only be coherently integrated with the question of value if the workings of natural causation by the past are relativized by a natural and, indeed, metaphysical teleology of value, one understood in process terms. Such a metaphysical teleology of value, I have argued, ultimately presupposes a divine individual that is internally related to and thus receives and integrates all that an evolving world produces. There is indeed grandeur in evolution so viewed because it is not just a long train to nowhere; rather, it is an ongoing creative process of generating various forms of life that all contribute varying degrees of beauty and worth to the world and to God, who receives and integrates all of the world into the everlasting, divine life.

BIBLIOGRAPHY

Allhoff, Fritz. "Evolutionary Ethics from Darwin to Moore." *History and Philosophy of Life Sciences* 25 (2003) 83–111.
Ansted, Daniel. "Interview: Michael Ruse on Evolution, Creationism, and Religion." January 4, 2013. Patheos. *Science on Religion*. http://www.patheos.com/blogs/sciencereligion/2013/01/interview-michael-ruse-on-evolution-creationism-and-religion/.
Anthes, Emily. "Coldblooded Does Not Mean Stupid." Science. *New York Times*, November 18, 2013. http://www.nytimes.com/2013/11/19/science/coldblooded-does-not-mean-stupid.html/.
Aristotle. *The Politics*. Translated by T. A. Sinclair. Revised and re-presented by Trevor J. Saunders. Penguin Classics. Harmondsworth, UK: Penguin, 1957.
Balfour, Arthur James. *The Foundations of Belief: Being Notes Introductory to the Study of Theology*. New York: Longmans Green, 1895.
Brown, Lesley, ed. *The New Shorter Oxford English Dictionary on Historical Principles*. 2 vols. Oxford: Clarendon, 1993.
Camus, Albert. *The Outsider*. Translated by Joseph Laredo. London: Penguin, 2000.
Carroll, Sean B. *Endless Forms Most Beautiful: The New Science of Evo Devo*. Illustrations by Jamie W. Carroll et al. New York: Norton, 2005.
Chang, Kenneth. "Size of Blast and Number of Injuries Are Seen as Rare for a Rock from Space." Space & Cosmos. *New York Times*, February 15, 2013. http://www.nytimes.com/2013/02/16/science/space/size-of-blast-and-number-of-injuries-are-seen-as-rare-for-a-rock-from-space.html?_r=0/.
Cobb, John B., Jr., ed. *Back to Darwin: A Richer Account of Evolution*. Grand Rapids: Eerdmans, 2008.
Conway Morris, Simon. *Life's Solution: Inevitable Humans in a Lonely Universe*. Cambridge: Cambridge University Press, 2003.
Coyne, Jerry A. *Why Evolution Is True*. New York: Viking, 2009.
Cunningham, Conor. *Darwin's Pious Idea: Why the Ultra-Darwinists and Creationists Both Get It Wrong*. Grand Rapids: Eerdmans, 2010.
Darwin, Charles. *The Autobiography of Charles Darwin*. Introduction by Brian Regal. Barnes & Noble Library of Essential Reading. New York: Barnes & Noble, 2005. (First published 1887.)

Bibliography

———. *The Descent of Man, and Selection in Relation to Sex*. Introduction by James Moore and Adrian Desmond. Penguin Classics. London: Penguin, 2004. Penguin is based on 2nd ed., 1879. (First published 1871.)

———. *The Expression of the Emotions in Man and Animals*. Edited by Joe Cain and Sharon Messenger with an introduction by Joe Cain. Penguin Classics. London: Penguin, 2009. (Based on 2nd ed., 1890; first published 1872.)

———. *The Origin of Species by Means of Natural Selection*. Edited and introduction by J. W. Burrow. Penguin Classics. London: Penguin, 1985. (Based on 1st ed., 1859.)

———. *The Variations of Animals and Plants under Domestication*. Edited by Paul H. Barrett and R. B. Freeman, with Peter Gautrey. 2 vols. *The Works of Charles Darwin* 19–20. Washington Square: New York University Press, 1988. 2nd ed. revised, 1875. (First published 1868.)

Davies, Paul Charles William. "Are We Alone in the Universe?" Opinion Pages. *New York Times*, November 18, 2013. http://www.nytimes.com/2013/11/19/opinion/are-we-alone-in-the-universe.html?_r=0/.

Davies, Paul Sheldon. *Subjects of the World: Darwin's Rhetoric and the Study of Agency in Nature*. Chicago: University of Chicago Press, 2009.

Dawkins, Richard. *The Selfish Gene*. 30th ann. ed. Oxford: Oxford University Press, 2006. (First published 1976.)

Dennett, Daniel C. *Darwin's Dangerous Idea: Evolution and the Meanings of Life*. New York: Simon & Schuster, 1995.

———. *Freedom Evolves*. New York: Viking, 2003.

Dewey, John. *A Common Faith*. Terry Lectures. New Haven: Yale University Press, 1934.

———. *Experience and Nature*. 1925. Reprinted, Kessinger Publishing's Rare Reprints. Whitefish, MT: Kessinger, 2012.

Fleming, Susan, dir. "A Murder of Crows." PBS *Nature*. October 24, 2010. http://www.pbs.org/wnet/nature/a-murder-of-crows-full-episode/5977/.

Gamwell, Franklin I. *Beyond Preference: Liberal Theories of Independent Associations*. Chicago: University of Chicago Press, 1984.

———. *Existence and the Good: Metaphysical Necessity in Morals and Politics*. Albany: State University of New York Press, 2011.

———. *Politics as a Christian Vocation: Faith and Democracy Today*. Cambridge: Cambridge University Press, 2005.

Gould, Stephen Jay. *Full House: The Spread of Excellence from Plato to Darwin*. New York: Three Rivers, 1996.

———. *Rocks of Ages: Science and Religion in the Fullness of Life*. New York: Ballantine Books, 1999.

Gray, Asa. *Darwinia: Essays and Reviews Pertaining to Darwinism*. 1876. Reprinted, Memphis: General Books, 2010.

Greene, Brian. *The Hidden Reality: Parallel Universes and the Deep Laws of the Cosmos*. New York: Vintage, 2011.

Hartshorne, Charles. *Born to Sing: An Interpretation and World Survey of Bird Song*. Bloomington: Indiana University Press, 1973.

———. *Creative Experiencing: A Philosophy of Freedom*. Edited by Donald Wayne Viney and Jincheol O. Albany: State University of New York Press, 2011.

———. *The Divine Relativity: A Social Conception of God*. The Terry Lectures. 1948. Reprinted, New Haven: Yale University Press, 1976.

Bibliography

———. *Insights and Oversights of Great Thinkers: An Evaluation of Western Philosophy.* SUNY Series in Systematic Philosophy. Albany: State University of New York Press, 1983.

———. *The Logic of Perfection.* LaSalle, IL: Open Court, 1962.

———. *A Natural Theology for Our Time.* Morse Lectures 1964. LaSalle, IL: Open Court, 1967.

———. *Whitehead's Philosophy: Selected Essays, 1935–1970.* Lincoln: University of Nebraska Press, 1972.

Hodge, Jonathan, and Gregory Radick, eds. *The Cambridge Companion to Darwin.* Series of Cambridge Companions. Cambridge: Cambridge University Press, 2003.

Hösle, Vittorio, and Christian Illies, eds. *Darwinism & Philosophy.* Notre Dame: University of Notre Dame Press, 2005.

Illies, Christian. "Darwin's A Priori Insight: The Structure and Status of the Principle of Natural Selection." In *Darwinism and Philosophy,* edited by Vittorio Hösle and Christian Illies, 58–82. Notre Dame: University of Notre Dame Press, 2005.

James, William. "Philosophical Conceptions and Practical Results." In *The Pragmatism Reader: From Peirce through the Present,* edited by Robert B. Talisse and Scott F. Aiken, 66–78. Princeton: Princeton University Press, 2011. (James originally delivered this paper in 1898 at the University of California Berkeley.)

Jonas, Hans. *The Phenomenon of Life: Toward a Philosophical Biology.* Foreword by Lawrence Vogel. 1966. Reprinted, Northwestern University Studies in Phenomenology & Existential Philosophy. Evanston, IL: Northwestern University Press, 2001.

Kolata, Gina. "Bits of Matter, Far from 'Junk,' Play Crucial Role." Science. *New York Times,* September 5, 2012. http://www.nytimes.com/2012/09/06/science/far-from-junk-dna-dark-matter-proves-crucial-to-health.html/.

Langer, Susanne K. *Philosophy in a New Key: A Study in the Symbolism of Reason, Rite, and Art.* 3rd ed. Cambridge: Harvard University Press, 1957.

Lowe, Victor. *Alfred North Whitehead: The Man and His Work.* Vol. 1, *1861–1910.* Baltimore: Johns Hopkins University Press, 1985.

Marcel, Gabriel. *Creative Fidelity.* Translated from the French and with an introduction by Robert Rosthal. New York: Farrar, Straus, 1964.

Mayr, Ernst. *Toward a New Philosophy of Biology: Observations of an Evolutionist.* Cambridge, MA: Belknap, 1988.

Meyer, William J. "Value and Conceptions of the Whole: The Views of Dewey, Nagel, and Gamwell." In *Deep Morality: Religious Metaphysics and Public Life,* edited by Kevin Schilbrack (under review for publication).

———. *Metaphysics and the Future of Theology: The Voice of Theology in Public Life.* Foreword by Schubert M. Ogden. Eugene, OR: Pickwick Publications, 2010.

Murdoch, Iris. *The Sovereignty of Good.* Studies in Ethics and the Philosophy of Religion. London: Routledge & Kegan Paul, 1970.

Nagel, Thomas. *Mind and Cosmos: Why the Materialist Neo-Darwinian Conception of Nature is Almost Certainly False.* New York: Oxford University Press, 2012.

Ogden, Schubert M. Metaphysical Notebooks: February 2011. Currently unpublished.

Overbye, Dennis. "A Case for Why We're Alone." Science Times section, *New York Times,* August, 4, 2015.

Peirce, Charles S. *Selected Writings (Values in a Universe of Chance).* Edited with an introduction and notes by Philip P. Wiener. 1958. Reprinted, New York: Dover, 1966.

Bibliography

Pleins, J. David. *The Evolving God: Charles Darwin on the Naturalness of Religion.* New York: Bloomsbury, 2013.

Pollan, Michael. *The Botany of Desire: A Plant's-Eye View of the World.* New York: Random House, 2001.

Popper, Karl. *The Logic of Scientific Discovery.* 1959. Reprinted, Routledge Classics. London: Routledge Classics, 2002. (First published in German in 1935.)

Ramachandran, V. S. *The Tell-Tale Brain: A Neuroscientist's Quest for What Makes Us Human.* New York: Norton, 2011.

Richards, Robert J. *The Romantic Conception of Life: Science and Philosophy in the Age of Goethe.* Science and Its Conceptual Foundations. Chicago: University of Chicago Press, 2002.

Ruse, Michael, and Edward O. Wilson. "The Evolution of Ethics." *New Scientist* 108/1478 (October 17, 1985) 50–52.

Schellenberg, J. L. *Evolutionary Religion.* Oxford: Oxford University Press, 2013.

Scruton, Roger. *The Soul of the World.* Princeton: Princeton University Press, 2014.

Sherburne, Donald W., ed. *A Key to Whitehead's "Process and Reality."* Chicago: University of Chicago Press, 1981. (First published 1966 by Macmillan.)

Singer, Irving. *Meaning in Life.* Vol. 1, *The Creation of Value.* 1996. Reprinted, Cambridge, MA: MIT Press, 2010.

Taha, Nadia. "Opting Out of Parenthood, with Finances in Mind." Your Money. *New York Times*, November 13, 2012. http://www.nytimes.com/2012/11/14/your-money/opting-out-of-parenthood-with-finances-in-mind.html?_r=0/.

Tegmark, Max. *Our Mathematical Universe: My Quest for the Ultimate Nature of Reality.* New York: Knopf, 2014.

Thorsen, Nils. "Longing for the End of All: Interview with Lars von Trier." http://www.melancholiathemovie.com/#_interview/.

Whitehead, Alfred North. *Adventures of Ideas.* 1933. Reprinted, New York: Free Press, 1967.

———. *The Concept of Nature: The Tarner Lectures delivered at Trinity College, Cambridge, 1919.* Lexington, KY: Philosopher's Stone, 2010.

———. *The Function of Reason.* Beacon Paperbacks 72. Boston: Beacon, 1958. First published 1929 by Princeton University Press.

———. *Process and Reality: An Essay in Cosmology.* Corrected edition by David Ray Griffin and Donald W. Sherburne. Gifford Lectures 1927–28. New York: Free Press, 1978. First published 1929.

———. *Science and the Modern World: Lowell Lectures.* 1925. Reprinted, New York: Macmillan, 1967.

Wilson, Edward O. *The Diversity of Life.* New York: Norton, 1992.

INDEX

Aristotle, 112, 114, 115

Cain, Joe, 67
Camus, Albert, 83, 104n49
Carroll, Sean, 38–39, 41
Cobb, John B., Jr., 6n10
Conway Morris, Simon, 35–36, 37–38
Coyne, Jerry, 2, 44n8, 55–56
creationism, 1, 5, 23

Darwin, Charles
    appreciation of beauty in nature, x, 1, 2, 71–73, 76
    art, literature, and music, 73–75, 76
    artificial selection, 17, 18, 19, 20, 25
    birds, their appreciation of beauty & novelty, ix, 6, 8, 12, 39, 45–49, 52, 53, 54, 55, 63, 72, 110, 113
    continuity of life, 40, 41–45, 47, 53, 63, 64
    death of daughter Annie, 74, 76, 89n17, 103n47, 113
    determinism and freedom, 10, 28–29
    domestication (domestic selection), 6, 12, 16, 17–21, 22, 23, 25, 26, 27, 28, 41, 47, 54, 110
    end of life on earth, 82–83
    evaluative assessment of nature, 1, 2, 7, 14, 71, 72, 83, 89, 110

God, question of, x, 4, 28–29
Gray, Asa, 7, 10, 28
human appreciation of beauty, ix, 39, 45
human capacity for self-conscious awareness, 12, 25, 64–66, 76, 78, 79, 80
human emotions, 42, 43, 67, 68, 69–71
Malthus, Thomas, 16, 17, 19, 23–26, 28
mechanistic view of nature, x, 9, 10, 14, 76
methodical selection, 17, 20
natural selection, 1, 2, 6, 18, 19, 20, 22, 24, 26–28, 35, 59, 116
quest for understanding, 7, 19, 65–68, 110
sexual selection, 12, 39, 45, 52–56, 116
unconscious selection, 17, 21–24, 27, 54
Darwin, Francis, 72–73, 76
Davies, Paul Sheldon, 14–15, 30–34
Dawkins, Richard, 3, 9, 39, 57
Dennett, Daniel, 2–3, 9, 10n15, 33, 35, 36–37, 56–62, 96, 110
Dewey, John, 83, 85
domestic selection, 16, 17–21, 22, 23, 25, 26, 27, 29, 30, 39, 41, 47, 54, 110

Ecclesiastes, 1, 2, 82
emotions as evaluations, 69–71

Index

existence with understanding (subjectivity), 65, 76–80

Gamwell, Franklin, 5n7, 65, 76–80, 83, 87, 88–89, 92, 105–6, 115
genetic fallacy, 34
Gould, Stephen Jay, 3–4, 4n4, 35, 36, 37, 38
Greene, Brian, 87–88

Hartshorne, Charles, 49–52, 60, 61, 77, 79, 83, 92, 100, 102–5, 106, 107–9, 110n56, 111–13

Illies, Christian, 26–27
intelligent design, 1, 2, 5

James, William, ix, x, 6n10, 89–92, 103, 104, 106
Jonas, Hans, 44–45, 50

Langer, Susanne, x, 14n20

Marcel, Gabriel, 89n17
Mayr, Ernst, 8–9, 29–30, 33, 34
mechanistic view of nature, x, 8–9, 10, 11, 13, 14, 30, 32, 57–60, 75, 76, 90, 91, 92, 93, 94, 97, 106, 110, 111, 115
Mill, John Stuart, 74
Murdoch, Iris, 61

Nagel, Thomas, 62n43, 79

Ogden, Schubert, 101n40

Paley, William, 6, 60, 94, 106, 110
Peirce, Charles, 30, 49, 61, 62, 104
Pleins, J. David, 4n5
Pollan, Michael, 10–12
Popper, Karl, 13–14
process (neoclassical) metaphysics
   actual occasions, 94–97, 111
   actual occasions vs. societies, 98–99
   composite individuals vs. compound individuals, 102–3

dipolarity, 107, 108–10
God and creation of value, 113–14
God and enduring value, 102, 114–15
God and temporality (time), 92, 107, 109
internal and external relations, 95–96, 97, 108
novelty and creativity, 95, 96–97
organic view of nature and reality, 92, 93–95, 96, 110
prehension, 94–95
process conception of God, 92, 106–10, 113–15
relatedness, fragmentary & transitory vs. holistic & everlasting, 101–2, 103, 107
relativity, 107–8
societies, nonpersonal and personal, 99–100
universe as compound, theistic arguments, 103–6
universe as societies of societies of societies, 95, 100–101
value, 111–13

reductionism, 41, 55, 56–62, 63, 73, 76
Richards, Robert, 10, 10n15
Ruse, Michael, 9, 57

Schellenberg, J.L., 83n4
Scruton, Roger, 4n4, 34n32
secularism (secularistic), 62, 82, 114–15
Sherburne, Donald W., 93
Singer, Irving, 83–88, 91, 92, 101–2, 105
Stout, Jeffrey, 62n43
subjectivity and transcendental analysis, 65, 76–80

Tegmark, Max, 82n2
teleology of design, ix, 5–6, 7, 8, 10, 20, 23, 27, 28, 29, 35, 38, 42,

Index

44, 63, 92, 106, 110, 111, 113, 114, 116
teleology of value, ix, 5, 6, 7, 8, 13, 14, 18, 20, 28, 29, 30, 38, 45, 58, 62, 63, 80, 84, 92, 93, 101, 110, 111, 113, 114, 116
three-dimensional reading of Darwin, x, 7, 8, 12, 13, 15, 41, 48, 49, 50, 55, 76, 116
Trier, Lars von, 81–82, 88
two-dimensional reading of Darwin, x, 7–8, 12, 13, 41, 44, 48, 55, 56, 76, 80, 116

Whitehead, Alfred N., ix, x, 5, 6n10, 8, 34, 34n32, 56, 62, 66, 71, 73, 75, 77, 79, 83, 92, 93–102, 103, 106, 109–10, 110n56, 111, 112, 113–15
Wilson, Edward O., 57
Wordsworth, William, 73, 75–76

www.ingramcontent.com/pod-product-compliance
Lightning Source LLC
Chambersburg PA
CBHW020856160426
43192CB00007B/954